HANDBOOK OF ELECTRONIC MATERIALS
Volume 1

HANDBOOK OF ELECTRONIC MATERIALS

Compiled by:

ELECTRONIC PROPERTIES INFORMATION CENTER

Hughes Aircraft Company
Culver City, California

Sponsored by:

AIR FORCE MATERIALS LABORATORY

Air Force Systems Command
Wright Patterson Air Force Base, Ohio

Volume 1:
OPTICAL MATERIALS PROPERTIES, 1971

Volume 2:
III-V SEMICONDUCTING COMPOUNDS, 1971

Volume 3:
SILICON NITRIDE FOR MICROELECTRONIC APPLICATIONS, PART I:
                                   PREPARATION AND PROPERTIES, 1971

In preparation:

Volume 4:
NIOBIUM ALLOYS AND COMPOUNDS

Volume 5:
GROUP IV SEMICONDUCTING COMPOUNDS

Volume 6:
SILICON NITRIDE FOR MICROELECTRONIC APPLICATIONS, PART II: APPLICATIONS

HANDBOOK OF ELECTRONIC MATERIALS
Volume 1

Optical Materials Properties

A. J. Moses

Electronic Properties Information Center
Hughes Aircraft Company, Culver City, California

IFI/PLENUM · NEW YORK-WASHINGTON-LONDON · 1971

This document has been approved for public release and sale;
its distribution is unlimited. Sponsored by : Air Force Materials
Laboratory, Wright-Patterson Air Force Base, Ohio.

Library of Congress Catalog Card Number 76-147312

ISBN-13: 978-1-4684-6161-9   e-ISBN-13: 978-1-4684-6159-6
DOI: 10.1007/978-1-4684-6159-6

FOREWORD

This report was prepared by Hughes Aircraft Company, Culver City, California under Contract Number F33615-70-C-1348. The work was administered under the direction of the Air Force Materials Laboratory, Air Force Systems Command, Wright-Patterson Air Force Base, Ohio, with Mr. B. Emrich, Project Engineer.

The Electronic Properties Information Center (EPIC) is a designated Information Analysis Center of the Department of Defense authorized to provide information to the entire DOD community. The purpose of the Center is to provide a highly competent source of information and data on the electronic, optical and magnetic properties of materials of value to the Department of Defense. Its major function is to evaluate, compile and publish the experimental data from the world's unclassified literature concerned with the properties of materials. All materials relevant to the field of electronics are within the scope of EPIC: insulators, semiconductors, metals, super-conductors, ferrites, ferroelectric, ferromagnetics, electroluminescents, thermionic emitters and optical materials. The Center's scope includes information on over 100 basic properties of materials; information generally regarded as being in the area of devices and/or circuitry is excluded.

CONTENTS

DATA SHEETS (Continued)

APPENDIX A

APPENDIX B

INTRODUCTION

The designer of optical systems must base the selection of optical materials on a knowledge of optical, physical, thermal and mechanical properties. Frequently, the selection of the materials must result from a trade-off analysis as no one material has an ideal set of properties. This compilation of Optical Materials Properties Data Sheets summarizes the vital properties, needed to select an optical material for use in the ultraviolet, visible and infrared regions of the electromagnetic spectrum. Forty-nine materials are described in this compilation and each sheet of two pages presents the properties of a particular material. The listed properties are:

Physical Properties: Density
                       Melting/Softening Temperature
                       Solubility in Water at Room Temperature

Thermal Properties: Linear Expansion Coefficient
                       Thermal Conductivity
                       Specific Heat

Mechanical Properties: Young's Modulus
                       Hardness

Optical Properties: Transmittance
                       Reflectance
                       Refractive Index
                       Extinction Coefficient or Absorption Coefficient
                       Dispersion Coefficient
                       Dispersion Equation

In some cases, even a thorough search of the literature did not yield data considered suitable for the data sheet and other useful plots are offered in place thereof. Most optical data are presented in graphical form as a function of wavelength. In some cases, wavenumber or photon energy are the independent variables, but conversion tables (Appendix A) will facilitate the use of these graphs. Optical terms are defined in Appendix B.

The computerized data retrieval system of the Electronic Properties Information Center, EPIC, provided most of the references for the data sheets (Figure 1). The EPIC system contains over 42,000 documents and covers 100 properties of more than 7,000 materials. The literature includes: (1) technical journals, (2) technical reports, (3) textbooks, (4) indexing and abstracting services, (5) theses and dissertations, (6) patents, (7) conference and symposium proceedings, (8) standards and specification documents, (9) vendor and trade literature, (10) bibliographies, and (11) private communications. This literature is in the following languages:

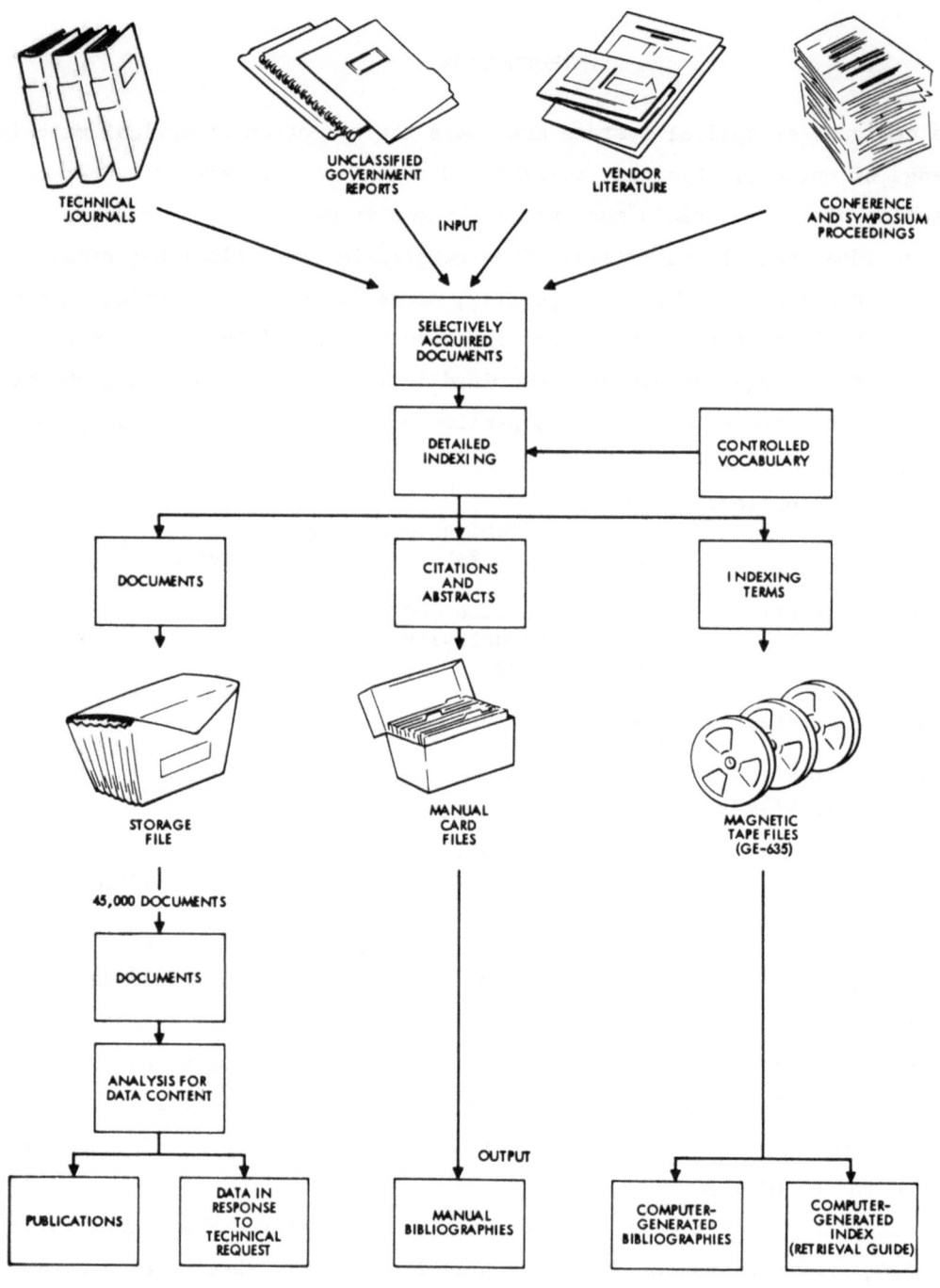

FIGURE 1

EPIC INFORMATION PROCESSING SYSTEM

English, German, French and Russian. The growth of the optical literature is illustrated by an over 200% increase in papers containing refractive index measurements over the seven year span from 1961 to 1968. The Electronic Properties Information Center, sponsored by the Air Force Materials Laboratory, includes optical materials properties in its coverage and its technical information analysis specialists can provide valuable aid to designers, scientists and engineers by assimilating and packaging pertinent information from this vast body of literature in the form of books and reports and the EPIC Technical Answering Service.

Technical Answering Service responds to inquiries ranging in complexity from simple requests for data point values to requests for comprehensive reviews of the literature. This service is available to U.S. Government agencies, their contractors, subcontractors, suppliers, and those in a position to support the defense effort. Inquiries may be directed to:

Electronic Properties Information Center
Hughes Aircraft Company
Bldg. 6: E-148
Centinela and Teale Streets
Culver City, California 90230
Telephone: (213) 391-0711, Ext. 6596.

The EPIC Bulletin, published quarterly, announces new publications and current activities of the Center. Users may request receipt of the Bulletin on a regular basis.

ALUMINA

OPTICAL MATERIALS PROPERTIES
DATA SHEET

MATERIAL: ALUMINA
(Sapphire)

INTRODUCTION: This data sheet provides information on single crystal synthetic sapphire. The optical data are for the ordinary ray.

PHYSICAL PROPERTIES, (298°K)

Density, (g/cm^3) ___3.98___

Melting/Softening
Temp. (°K) ___2303___

Solubility in Water,
(g./100 g. H$_2$O) ___9.8 x 10^{-5}___

MECHANICAL PROPERTIES*, (298°K)

Young's Modulus, (psi) ___50 x 10^6___

Hardness, (Knoop) ___2000, (1000 g. indenter)___

THERMAL PROPERTIES*, (298°K)

Linear Expansion
Coeff., (°K)$^{-1}$ ___5.8 x 10^{-6}___

Thermal Conductivity
(10^{-4} cal/(cm sec °K) ___0.10___

Specific Heat, (cal/g)/°K ___0.18___

*60° orientation to optic axis

OPTICAL PROPERTIES, (298°K)

Dispersion Equation

$$n^2 - 1 = \sum_i \frac{A_i \lambda^2}{\lambda^2 - \lambda_t^2}$$

λ_1 = 0.06144821 λ_1^2 = 0.00377588 A_1 = 1.023798

λ_2 = 0.1106997 λ_2^2 = 0.0122544 A_2 = 1.058264

λ_3 = 17.92656 λ_3^2 = 321.3616 A_3 = 5.280792

(0.26 - 5.6 μ; λ in μ.)

Transmission Region, (External
Transmittance ≥ 10% with ___2.0___ mm.
thickness) ___0.15 - 6.5 μ___

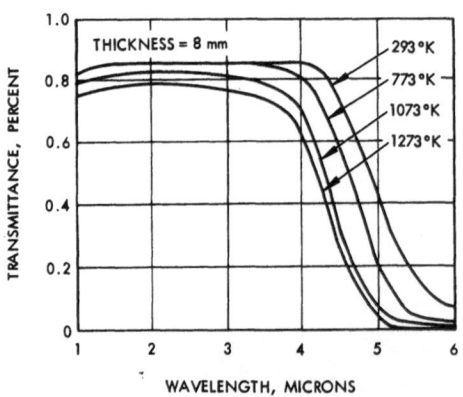

(REF. 1)

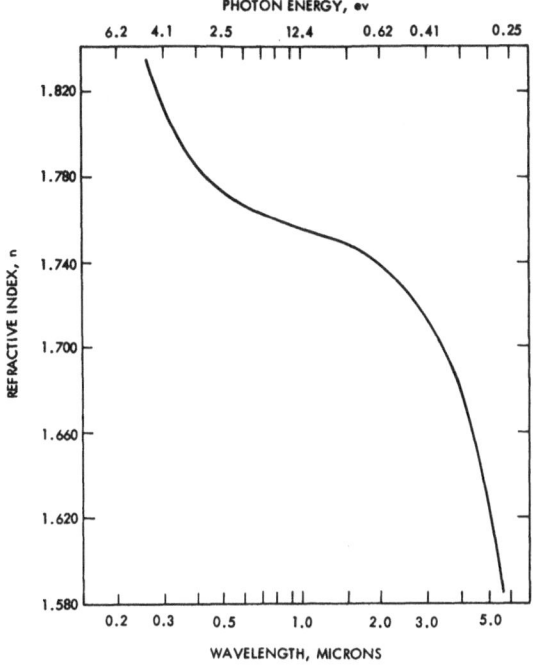

(REF. 2)

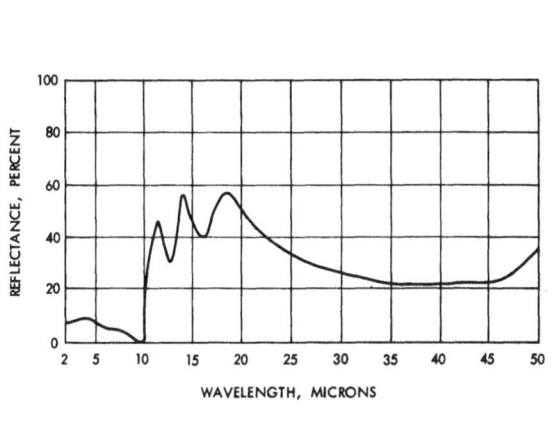

(REF. 3)

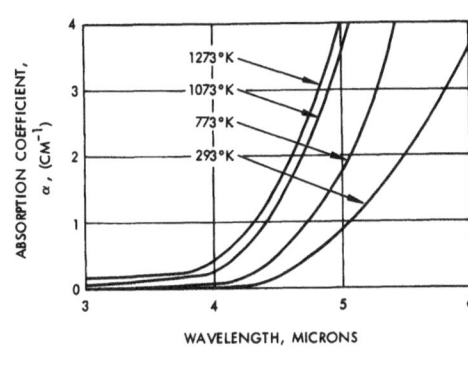

(REF. 1)

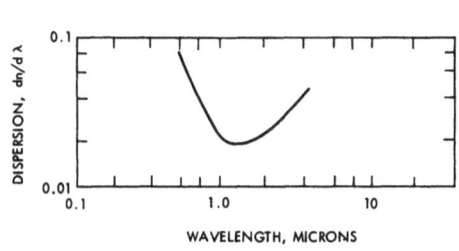

(REF. 4)

REFERENCES:

1. U. P. Oppenheim, and U. Even, J. Opt. Soc. Am., <u>52</u>, 1078-9, (1962).

2. D. E. McCarthy, Appl. Optics, <u>2</u>, 591-595, (1963).

3. I. H. Malitson, et al, J. Opt. Soc. Am., <u>48</u>, 72-73, (1958).

4. W. L. Wolfe, Ed., "Handbook of Military Infrared Technology," U.S. Govt. Printing Office, Washington, (1965).

ALUMINUM

OPTICAL MATERIALS PROPERTIES
DATA SHEET

MATERIAL: ALUMINUM
(Film)

INTRODUCTION: This data sheet presents data for aluminum film, formed by evaporation. The optical data are compromised by uncertainty concerning the crystallinity of the film.

PHYSICAL PROPERTIES, (298°K)

Density, (g/cm^3) _____ 2.70 _____

Melting/Softening
Temp. (°K) _____ 933 _____

Solubility in Water,
(g./100 g. H$_2$O) _____ 0.000 _____

MECHANICAL PROPERTIES, (298°K)

Young's Modulus, (psi) _____ 7.06 x 10^5

Hardness _____ ~25 Brinnell _____

THERMAL PROPERTIES, (298°K)

Linear Expansion
Coeff. (°K)$^{-1}$ _____ 23.5 x 10^{-6} _____

Thermal Conductivity
(10^{-4}cal/(cm sec °K) _____ 0.504 _____

Specific Heat, (cal/g)/°K _0.214_

OPTICAL PROPERTIES, (298°K)

Dispersion Equation: _____ not available _____

Transmission Region,
(External Transmittance
≥10% with _____ mm.
thickness) _____ not available _____

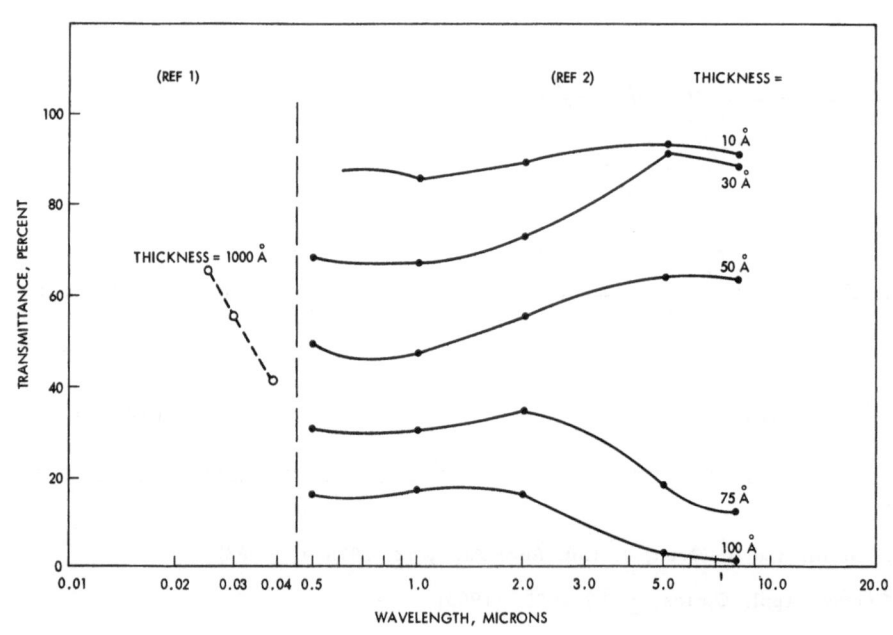

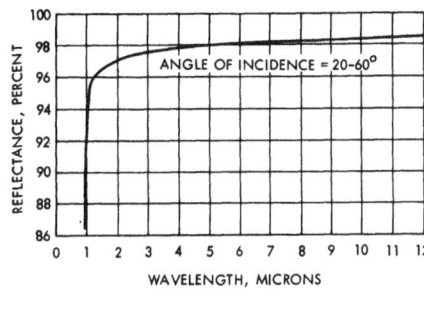

(REF. 3)

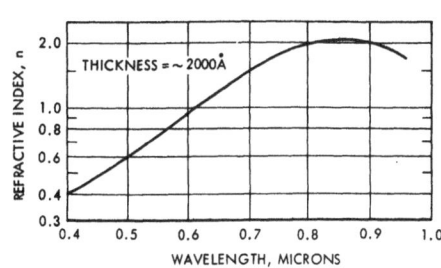

(REF. 4)

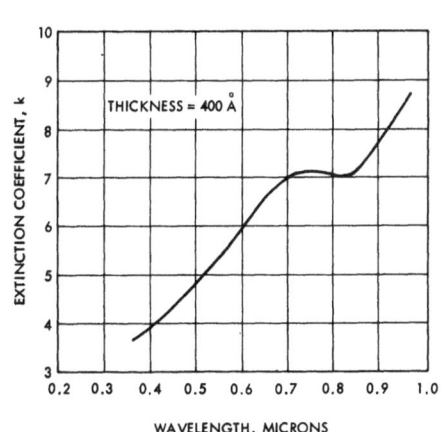

(REF. 5)

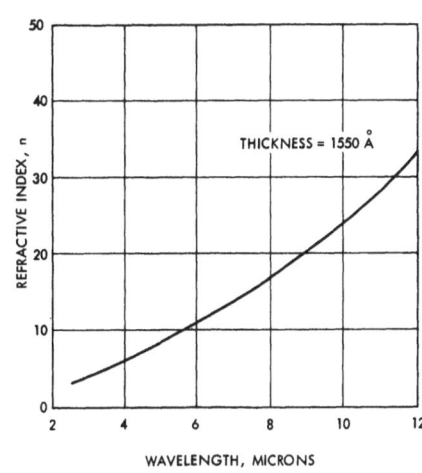

(REF. 6)

REFERENCES:

1. O. P. Rustgi, J. Opt. Soc. Am., 55, 630-634, (1965).

2. H. Mayer, "Physik duenner Schichten", Pt. I., Wiss. Verlagsgesell., Stuttgart, (1950).

3. D. M. Gates, et al., J. Opt. Soc. Am., 48, 88-89, (1958).

4. L. G. Schulz, and F. R. Tangherlini, J. Opt. Soc. Am., 44, 362-368, (1954).

5. L. G. Schulz, J. Opt. Soc. Am., 44, 357-362, (1954).

6. J. R. Beattie, Phil. Mag., 46, 235-245, (1955).

AMMONIUM DIHYDROGEN PHOSPHATE

OPTICAL MATERIALS PROPERTIES
DATA SHEET

MATERIAL: AMMONIUM
DIHYDROGEN
PHOSPHATE
(ADP)

INTRODUCTION: This data sheet contains information on single crystal ammonium dihydrogen phosphate.

PHYSICAL PROPERTIES, (298°K)

Density, (g/cm^3)   1.803

Melting/Softening
Temp. ($^{\circ}$K)   526

Solubility in Water,
(g./100 g. H$_2$O)   22.7 (273°K)

MECHANICAL PROPERTIES, (298°K)

Young's Modulus, (psi)   Not available

Hardness, (Knoop)   Not available

THERMAL PROPERTIES, (298°K)

Linear Expansion
Coeff., ($^{\circ}$K)$^{-1}$ 39×10^{-6} (a axis)
                                 1.9×10^{-6} (c axis)

Thermal Conductivity   17 (c axis);
(10^{-4} cal/(cm sec $^{\circ}$K)   30 (a axis)

Specific Heat,
(cal/g)/$^{\circ}$K   0.29

OPTICAL PROPERTIES, (298°K)

Dispersion Equation   Not available

Transmission Region, (External
Transmittance $\geq$10% with 2.0 mm.
thickness)   0.13 - 1.7μ

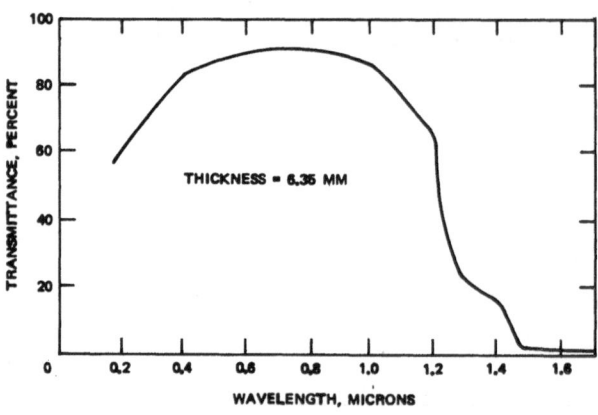

THICKNESS = 6.35 MM

(REF. 1)

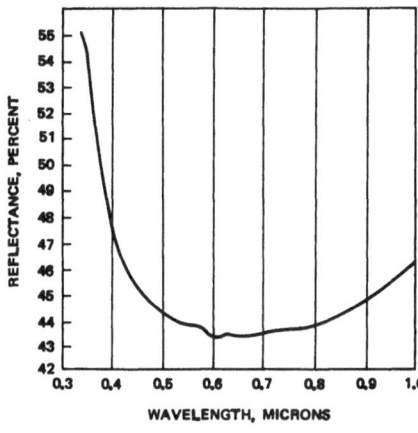

(REF. 2)

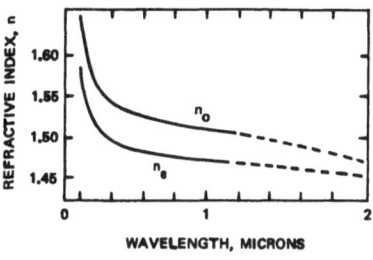

(REF. 3)

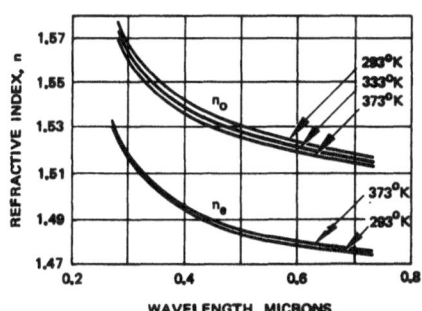

(REF. 4)

REFERENCES:

1. G.D. Burnett, Electronic Industries, <u>21</u>, 90-95, (1962).

2. F. Vratny and J.J. Kokalas, Appl. Spectroscopy, <u>16</u>, 176-184, (1962).

3. F. Zernike, Jr., J. Opt. Soc. Am., <u>54</u>, 1215-1220, (1964).

4. V.N. Vishneskii, et al, Opt. Spec., (English Transl.), <u>18</u>, 468-469, (1965).

ARSENIC TRISULFIDE

OPTICAL MATERIALS PROPERTIES
DATA SHEET

MATERIAL: __ARSENIC TRISULFIDE (Glass)__

INTRODUCTION: This data sheet presents data for arsenic trisulfide glass.

PHYSICAL PROPERTIES, (298°K)

Density, (g/cm^3) 3.198

Melting/Softening
Temp. (°K) 483

Solubility in Water,
(g./100 g. H$_2$O) ~5 x 10^{-5} (291°K)

MECHANICAL PROPERTIES, (298°K)

Young's Modulus, (psi) 2.3 x 10^6

Hardness, (Knoop) 109 (100 g.)

THERMAL PROPERTIES, (298°K)

Linear Expansion
Coeff. (°K)$^{-1}$ 24.6 x 10^{-6}

Thermal Conductivity
(10 cal/(cm sec °K) 4.0 (313°K)

Specific Heat, (cal/g)/°K not available

OPTICAL PROPERTIES, (298°K)

Dispersion Equation:

$$n^2 - 1 = \sum_{i=1}^{5} \frac{K_i \lambda^2}{\lambda^2 - \lambda_i^2}$$

i	λ_i^2	K_i
1	0.0225	1.8983678
2	0.0625	1.9222979
3	0.1225	0.8765134
4	0.2025	0.1188704
5	750.	0.9569903

Transmission Region,
(External Transmittance
10% with 2.0 mm.
thickness) 0.6 - 13μ

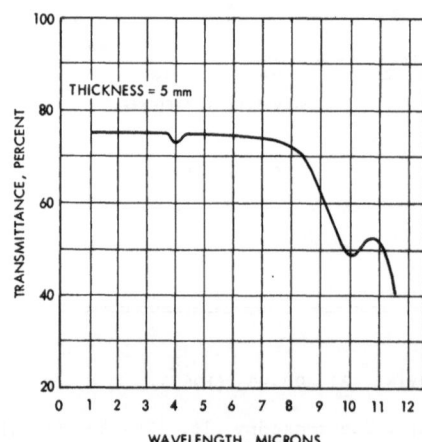

(REF. 1)

10

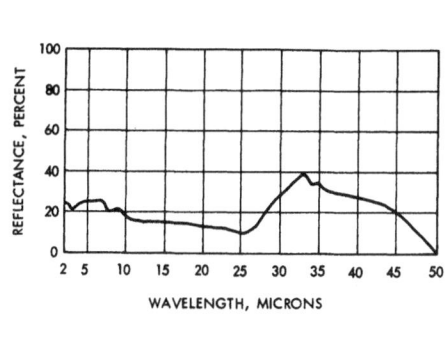

(REF. 2)

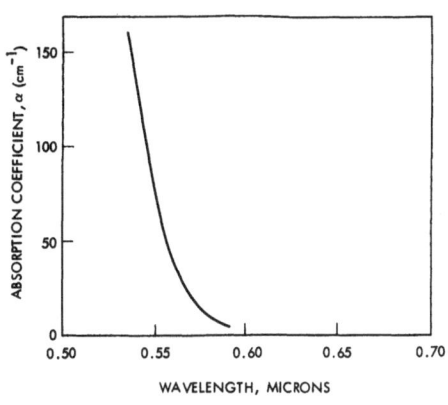

(REF. 4)

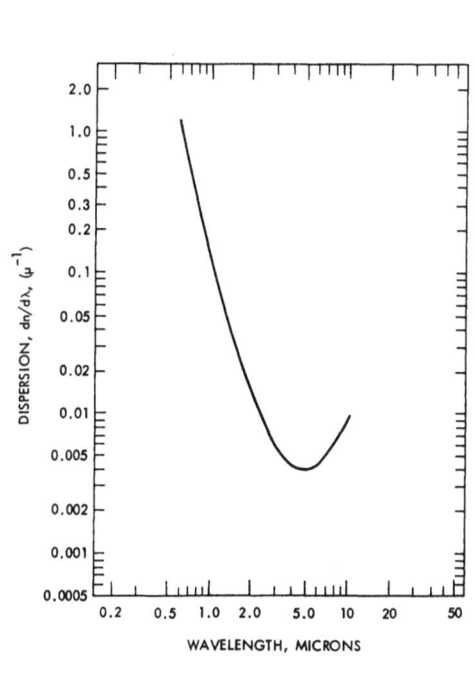

(REF. 3)

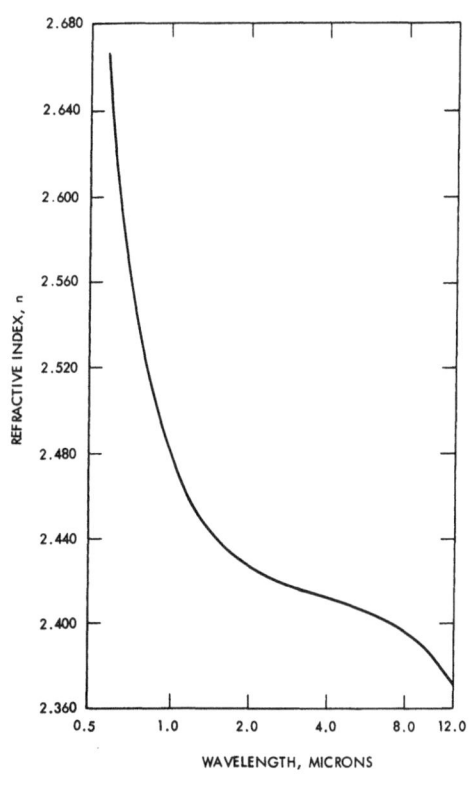

(REF. 3)

REFERENCES:

1. S. S. Ballard, et al., "Optical Materials for Infrared Instrumentation," AD 217 367, (1959).

2. D. E. McCarthy, Appl. Optics, 2, 591-595, (1963).

3. W. S. Rodney, et al., J. Opt. Soc. Am., 48, 633-636, (1958).

4. G. Getov, et al., Phys. Stat. Sol., 21, K87-K89, (1967).

BARIUM FLUORIDE

OPTICAL MATERIALS PROPERTIES
DATA SHEET

MATERIAL: BARIUM FLUORIDE

INTRODUCTION: This data sheet provides information for synthetic single crystal barium fluoride.

PHYSICAL PROPERTIES, (298°K)

Density, (g/cm^3) _____ 4.83

Melting/Softening
Temp. (°K) _____ 1553

Solubility in Water,
(g./100 g. H$_2$O) _____ 0.16

MECHANICAL PROPERTIES, (298°K)

Young's Modulus, (psi) _____ 7.7 x 10^{-6}

Hardness, (Knoop) _____ 82, (500 g.)

THERMAL PROPERTIES, (298°K)

Linear Expansion
Coeff. (°K)$^{-1}$ _____ 18.4 x 10^{-6}

Thermal Conducitivity
(10^{-4}cal/(cm sec °K) _____ 280

Specific Heat, (cal/g)/°K _not available_

OPTICAL PROPERTIES, (298°K)

Dispersion Equation:

$$n^2 - 1 = \frac{0.643356\lambda^2}{\lambda^2 - (0.057789)^2} + \frac{0.50676\lambda^2}{\lambda^2 - (0.10968)^2}$$

$$+ \frac{3.8261\lambda^2}{\lambda^2 - (46.3864)^2}$$

Transmission Region,
(External Transmittance
≥ 10% with _2.0_ mm.
thickness) _____ 0.25 - 15 μ

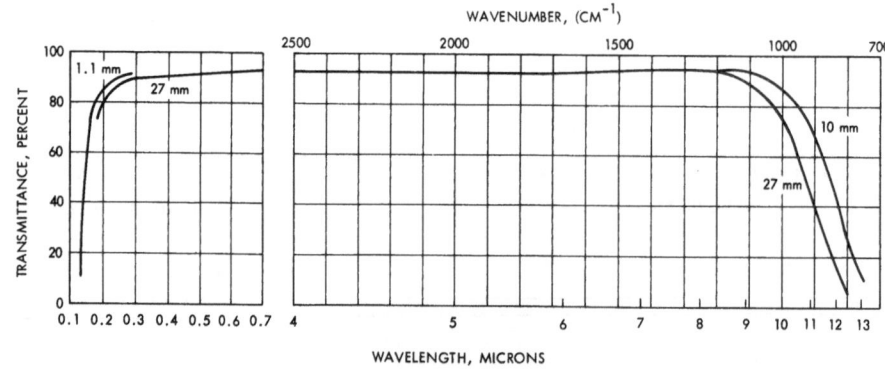

(REF. 1)

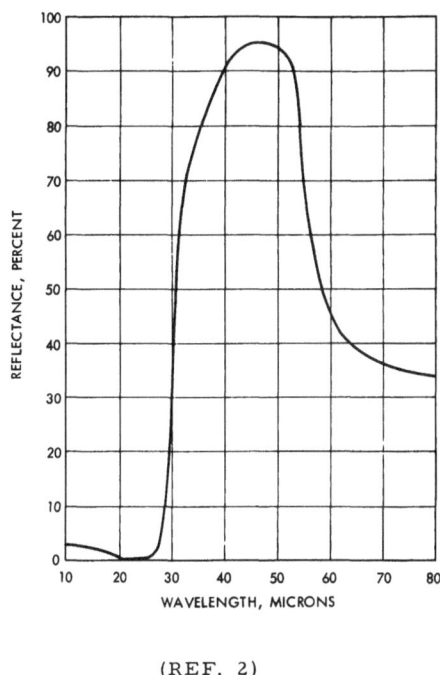

(REF. 2)

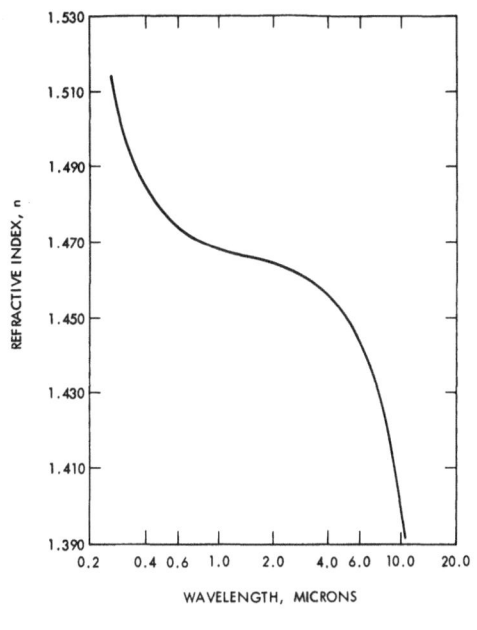

(REF. 3)

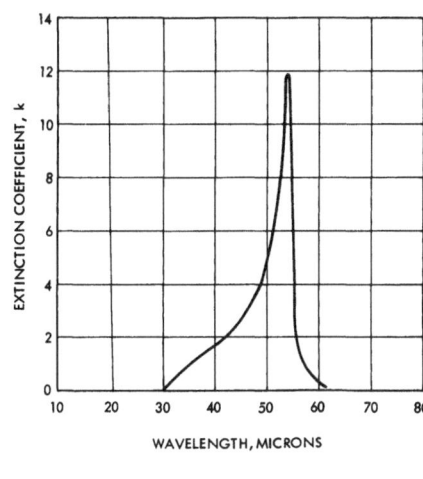

(REF. 2)

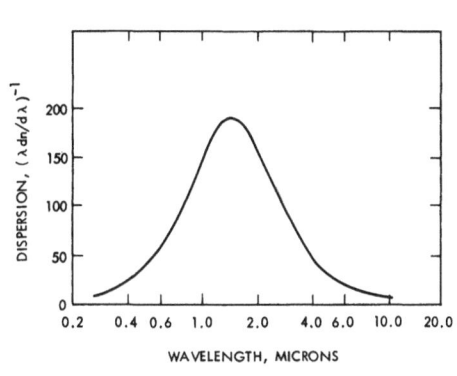

(REF. 3)

REFERENCES:

1. A. Smakula, et al., "Harshaw Optical Crystals", Harshaw Chemical Co., Cleveland, (1967).

2. W. Kaiser, et al., Phys. Rev., 127, 1950-1954, (1962).

3. I. H. Malitson, J. Opt. Soc. Am., 54, 628 - 632, (1964).

BERYLLIUM OXIDE

OPTICAL MATERIALS PROPERTIES
DATA SHEET

MATERIAL: BERYLLIUM
OXIDE

INTRODUCTION: This data sheet contains information for single crystal beryllium oxide.

PHYSICAL PROPERTIES, (298°K)

Density, (g/cm^3)   3.00

Melting/Softening
Temp. (°K)   2823

Solubility in Water,
(g./100 g. H$_2$O)   <0.01

MECHANICAL PROPERTIES, (298°K)

Young's Modulus, (psi)   53 x 10^6

Hardness, (Knoop)   917 - 1300

THERMAL PROPERTIES, (298°K)

Linear Expansion
Coeff., (°K)$^{-1}$ 5.5 x 10^{-6}

Thermal Conductivity
(10^{-4} cal/(cm sec °K)   520

Specific Heat,
(cal/g)/°K   0.24

OPTICAL PROPERTIES, (298°K)

Dispersion Equation   Not available

Transmission Region, (External
Transmittance ≥10% with_____mm.
thickness)   Not available

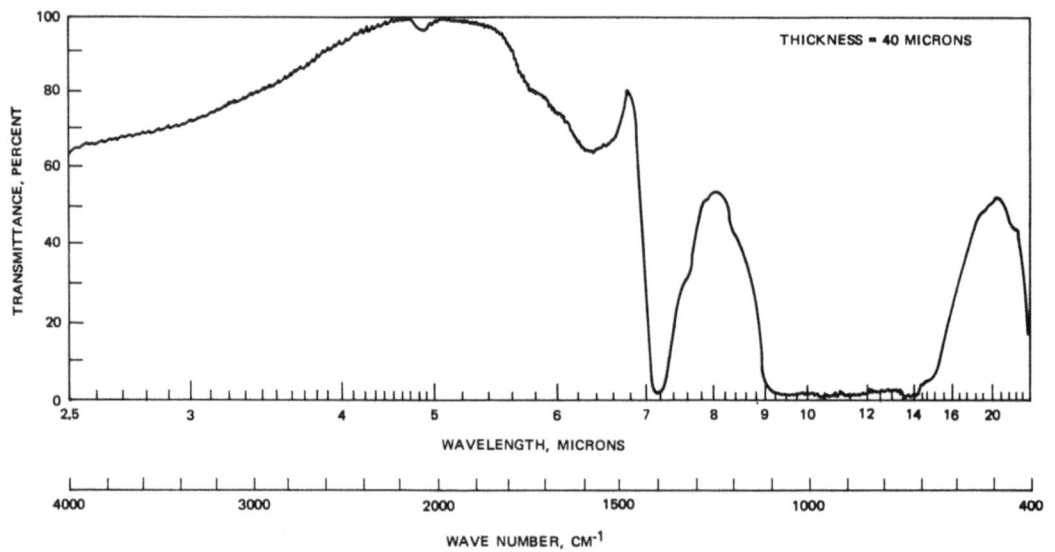

(REF. 1)

14

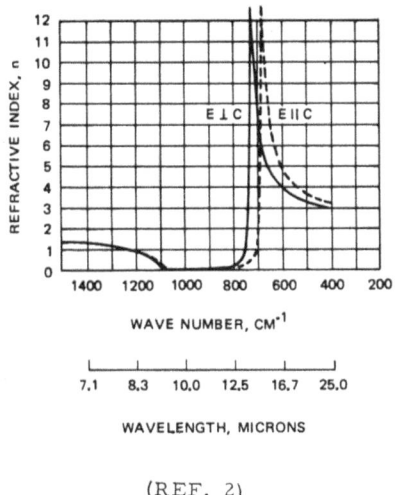

(REF. 2)

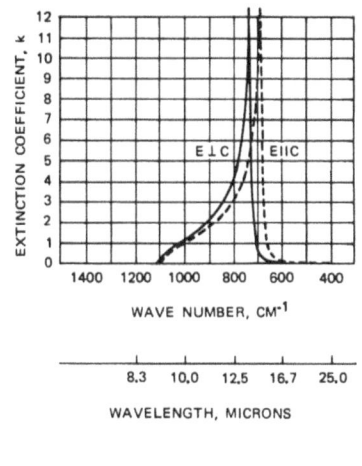

(REF. 2)

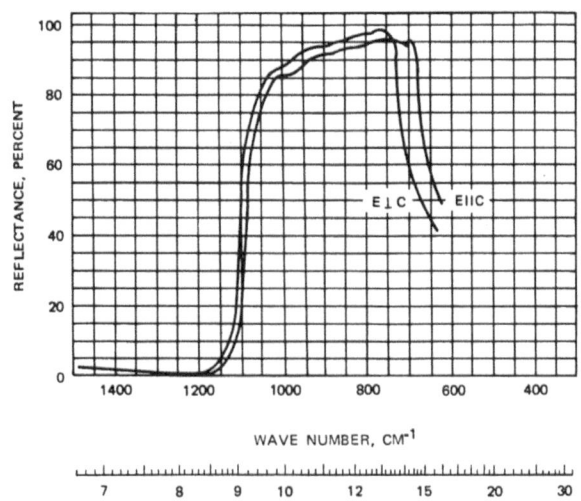

(REF. 2)

REFERENCES:

1. R.J. Morrow and H.W. Newkirk, Rev. Int. Hautes Temp. Refractaires, 6, No. 2, 99-104, (1969).

2. E. Loh, Phys. Rev., 166, 673-678, (1968).

CADMIUM SULFIDE

OPTICAL MATERIALS PROPERTIES
DATA SHEET

MATERIAL: CADMIUM SULFIDE

INTRODUCTION: This data sheet summarizes the properties of bulk, hexagonal single crystal cadmium sulfide.

PHYSICAL PROPERTIES, (298°K)

Density, (g/cm^3) ___4.82___

Melting/Softening
Temp. (°K) ___1253 (subl. at 1 atm.)___

Solubility in Water,
(g./100 g. H_2O)___1.3×10^{-4} (291°K)___

MECHANICAL PROPERTIES, (298°K)

Young's Modulus, (psi) ___not available___

Hardness, (Knoop) ___121, (100 g)___
___(11 to C axis)___

THERMAL PROPERTIES, (298°K)

Linear Expansion
Coeff. $(°K)^{-1}$ ___5×10^{-6} (⊥C axis)___
___3.5×10^{-6} (11 C axis)___

Thermal Conductivity
$(10^{-4}$ cal/(cm sec °K)___380___

Specific Heat, (cal/g)/°K ___0.088 (273°K)___

OPTICAL PROPERTIES, (298°K)

Dispersion Equation

$$n_o^2 = 5.235 + \frac{1.819 \times 10^7}{\lambda^2 - 1.651 \times 10^7}$$

$$n_e^2 = 5.239 + \frac{2.076 \times 10^7}{\lambda^2 - 1.651 \times 10^7}$$

Transmission Region, (External Transmittance ≥ 10% with ___2.0___ mm. thickness) ___$0.5 - 16\mu$___

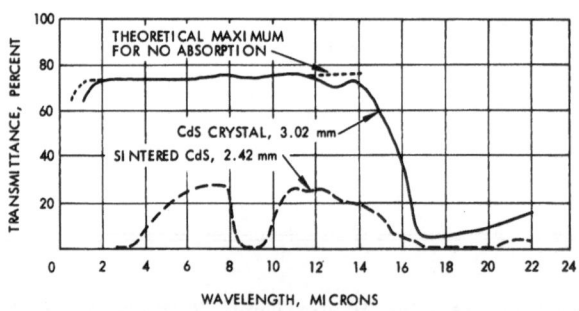

(REF. 1)

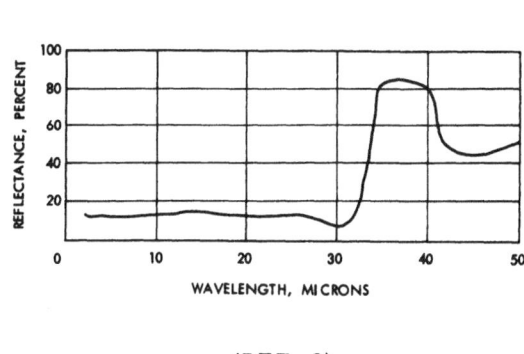

(REF. 2)

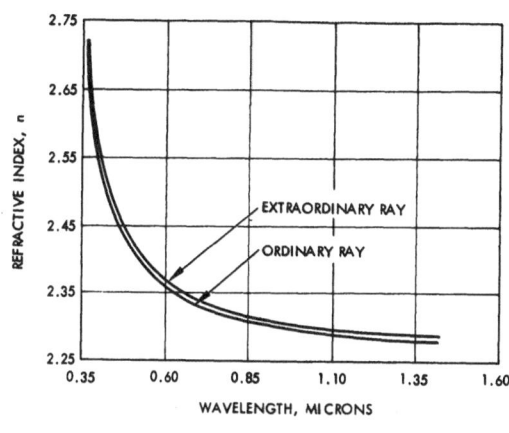

(REF. 3)

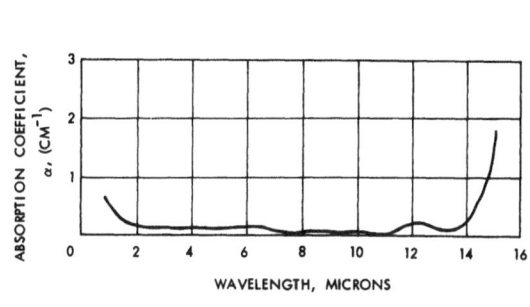

(REF. 1)

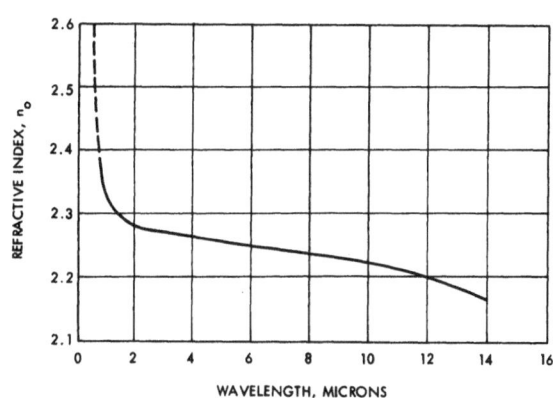

(REF. 1)

REFERENCES:

1. A. B. Francis, and A. I. Carlson, J. Opt. Soc. Am., <u>50</u>, 118-121, (1960).

2. D. E. McCarthy, Applied Optics, <u>7</u>, 1997-2000, (1968).

3. T. M. Bieniewski, and S. J. Czyzak, J. Opt. Soc. Am., <u>53</u>, 649-497, (1963).

CADMIUM TELLURIDE

OPTICAL MATERIALS PROPERTIES   MATERIAL: <u>CADMIUM</u>
                                                 <u>TELLURIDE</u>

INTRODUCTION: This data sheet presents data for hot-pressed polycrystalline cadmium telluride.

PHYSICAL PROPERTIES, (298°K)

Density, (g/cm^3) _____ 5.85

Melting/Softening
Temp. (°K) _____ 1318

Solubility in Water,
(g./100 g. H$_2$O) _____ not available

MECHANICAL PROPERTIES, (298°K)

Young's Modulus, (psi) _____ 5.3 x 10^6

Hardness, (Knoop) _____ 45

THERMAL PROPERTIES, (298°K)

Linear Expansion
Coeff. (°K) _____ 5.5 x 10^{-6}

Thermal Conductivity
(10^{-4} cal/(cm sec °K) _____ 98

Specific Heat, (cal/g)/°K _____ not available

OPTICAL PROPERTIES, (298°K)

Dispersion Equation: _____ not available

Transmission Region,
(External Transmittance
≧10% with _____ 2.0 _ mm.
thickness) _____ 0.9 - 16µ

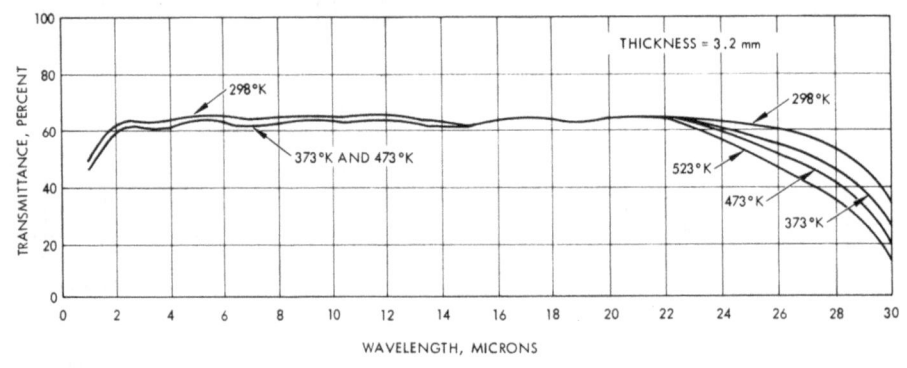

(REF. 1)

18

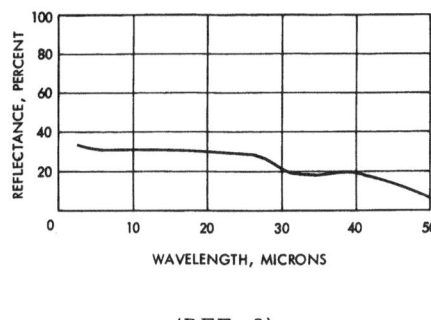

(REF. 2)

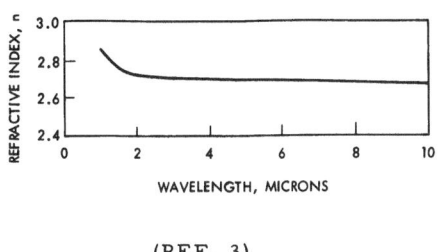

(REF. 3)

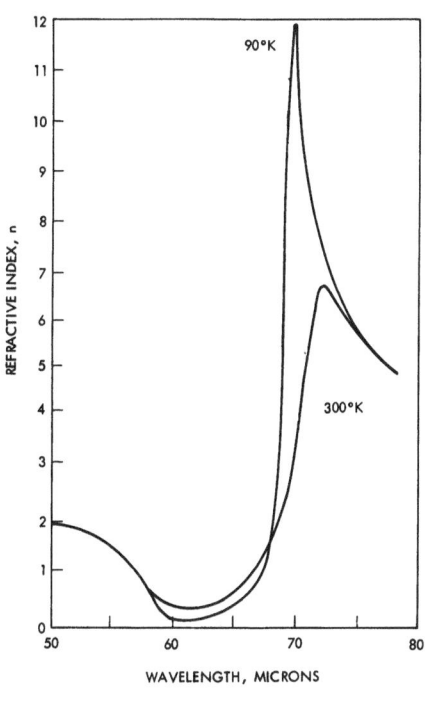

(REF. 4)

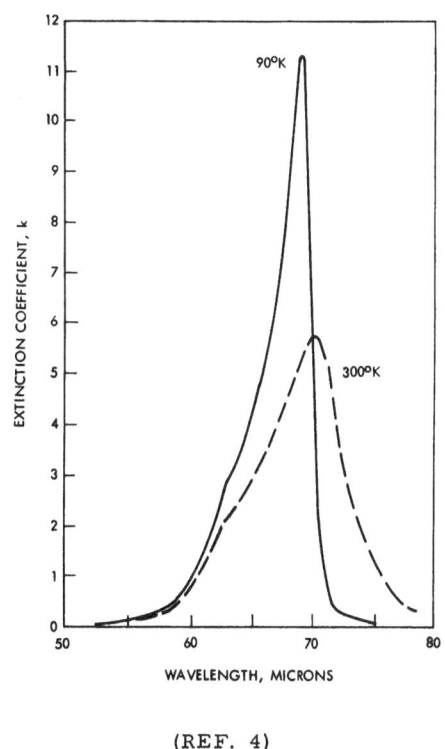

(REF. 4)

REFERENCES:

1.   L. S. Ladd, Infrared Physics, 6, 145-151, (1966).

2.   D. E. McCarthy, Appl. Optics, 7, 1997-2000, (1968).

3.   Eastman Kodak Company, Kodak IRTRAN-6 Material Data Sheet.

4.   A. Mitsuishi, J. Phys. Soc. Japan, 16, 533-537, (1961).

CALCIUM CARBONATE

OPTICAL MATERIALS PROPERTIES
DATA SHEET

MATERIAL: <u>CALCIUM</u>
<u>CARBONATE</u>
<u>(Calcite)</u>

INTRODUCTION: <u>This data sheet provides data for single crystal calcium carbonate.</u>

PHYSICAL PROPERTIES, (298°K)

Density, (g/cm^3) ___2.71___

Melting/Softening
Temp. (°K) ___1167 (Dissocn)___

Solubility in Water,
(g./100 g. H_2O) ___0.0014___

MECHANICAL PROPERTIES*, (298°K)

Young's Modulus, (psi) ___10.5, 12.8___

Hardness, (Mohs) ___3___

THERMAL PROPERTIES*, (273°K)

Linear Expansion
Coeff., (°K)$^{-1}$ ___25, (-) 5.8___

Thermal Conductivity
(10^{-4} cal/(cm sec °K) ___132, 111___

Specific Heat,
(cal/g)/°K ___0.203___

*E ‖ C and E⊥C respectively

OPTICAL PROPERTIES, (298°K)

Dispersion Equation ___Not available___

Transmission Region, (External
Transmittance ≥10% with ___2.0___ mm.
thickness) ___0.2 - 5.5µ___

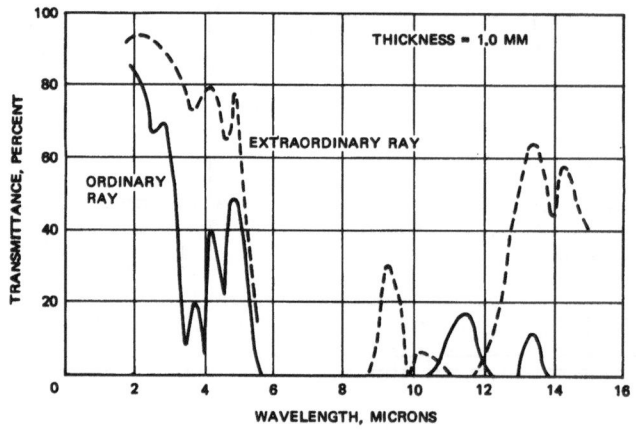

(REF. 1)

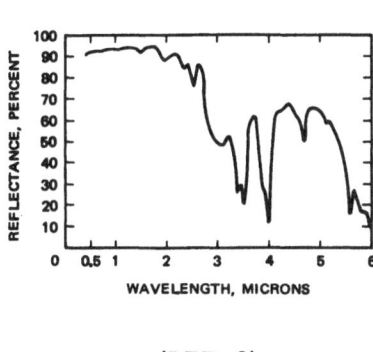

(REF. 2)

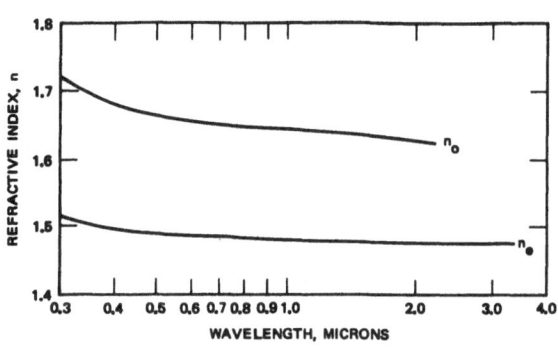

(REF. 1)

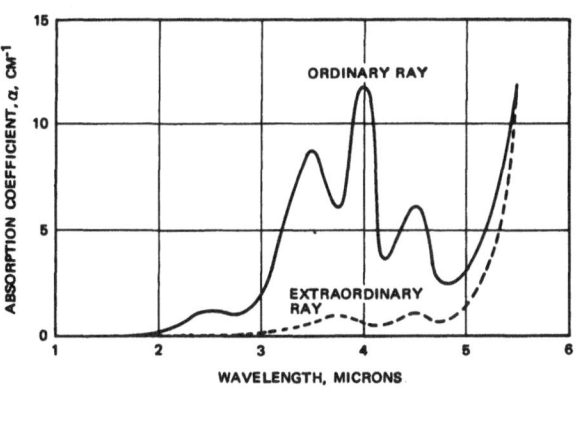

(REF. 3)

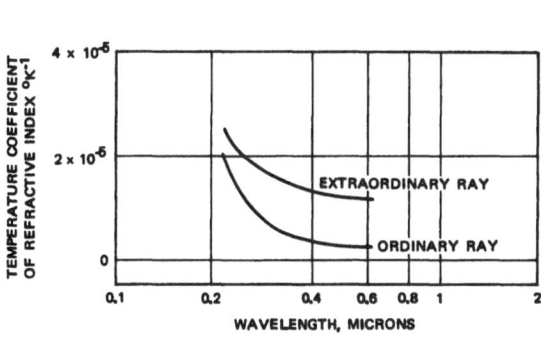

(REF. 3)

REFERENCES:

1. S.S. Ballard, et al., "Optical Materials for Infrared Instrumentation," Report No. AD 217367, (1959).

2. W.A. Hovis, Jr., Appl. Optics, 5, 245-248, (1966).

3. A. Smakula, Opt. Acta, 9, 205-222, (1962).

CALCIUM FLUORIDE

OPTICAL MATERIALS PROPERTIES
DATA SHEET

MATERIAL: <u>CALCIUM</u>
<u>FLUORIDE</u>

INTRODUCTION: This data sheet presents information for single crystal calcium fluoride.

PHYSICAL PROPERTIES, (298°K)

Density, (g/cm^3) ___3.179___

Melting/Softening
Temp. (°K) ___1613___

Solubility in Water,
(g./100 g. H$_2$O) ___0.0017___

MECHANICAL PROPERTIES, (298°K)

Young's Modulus, (psi) ___11.0 x 10^{-6}___

Hardness, (Knoop) ___158.3 (500 g)___

THERMAL PROPERTIES, (298°K)

Linear Expansion
Coeff. (°K)$^{-1}$ ___24 x 10^{-6}___

Thermal Conductivity
(10^{-4} cal/(cm sec °K) ___232 (309°K)___

Specific Heat, (cal/g)/°K ___0.204 (273°K)___

OPTICAL PROPERTIES, (298°K)

Dispersion Equation:

$$n^2 - 1 = \frac{0.5675888\lambda^2}{\lambda^2 - (0.050263605)^2} + \frac{0.4710914\lambda^2}{\lambda^2 - (0.1003909)^2}$$

$$+ \frac{3.8484723\lambda^2}{\lambda^2 - (34.649040)^2}$$

Transmission Region,
(External Transmittance
≳ 10% with ___2.0___ mm.
thickness) ___0.13 - 12 μ___

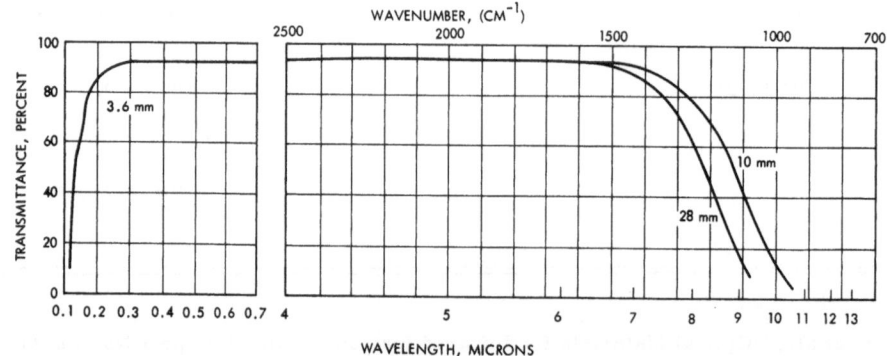

(REF. 1)

22

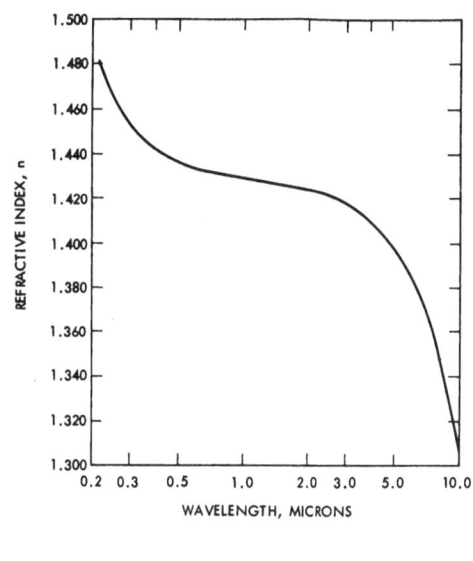

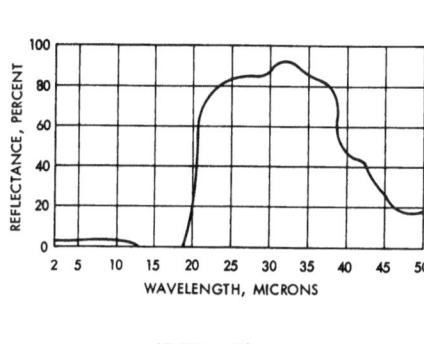

(REF. 2)

(REF. 3)

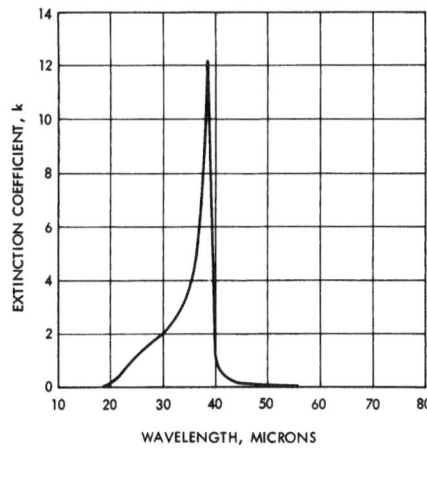

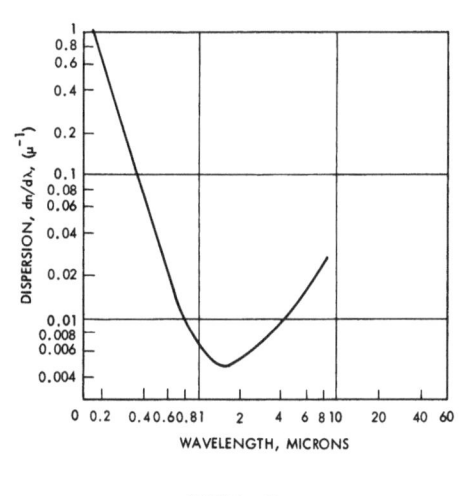

(REF. 4)

(REF. 5)

REFERENCES:

1. A. Smakula, et al., "Harshaw Optical Crystals," Harshaw Chemical Co., Cleveland, (1967).

2. D. E. McCarthy, Appl. Optics, 2, 591-595, (1963).

3. I. H. Malitson, Appl. Optics, 2, 1103-1107, (1963).

4. W. Kaiser, et al., Phys. Rev., 127, 1950-1954, (1962).

5. A. Smakula, Optica Acta, 9, 205-222, (1962).

CESIUM BROMIDE

OPTICAL MATERIALS PROPERTIES
DATA SHEET

MATERIAL: <u>CESIUM</u>
<u>BROMIDE</u>

INTRODUCTION: <u>This data sheet presents information on single crystal cesium bromide.</u>

PHYSICAL PROPERTIES, (298°K)

Density, (g/cm^3) ___4.44___

Melting/Softening
Temp. (°K) ___909___

Solubility in Water,
(g./100 g. H$_2$O) ___124___

MECHANICAL PROPERTIES, (298°K)

Young's Modulus, (psi) ___2.3 x 10^6___

Hardness, (Knoop) ___19.5 (200 g.)___

THERMAL PROPERTIES, (298°K)

Linear Expansion
Coeff., (°K)$^{-1}$ ___47.9 x 10^{-6}___

Thermal Conductivity
(10^{-4} cal/(cm sec °K) ___0.23___

Specific Heat, (cal/g)/°K ___0.063___

OPTICAL PROPERTIES, (298°K)

Dispersion Equation:

$$n^2 = 5.640752 - 0.000003338\lambda^2 + \frac{0.0018612}{\lambda^2}$$

$$+ \frac{41110.49}{\lambda^2 - 14390.4} + \frac{0.0290764}{\lambda^2 - 0.024964}$$

Transmission Region,
(External Transmittance
≥ 10% with ___2.0___ mm.
thickness) ___0.3 - 55µ___

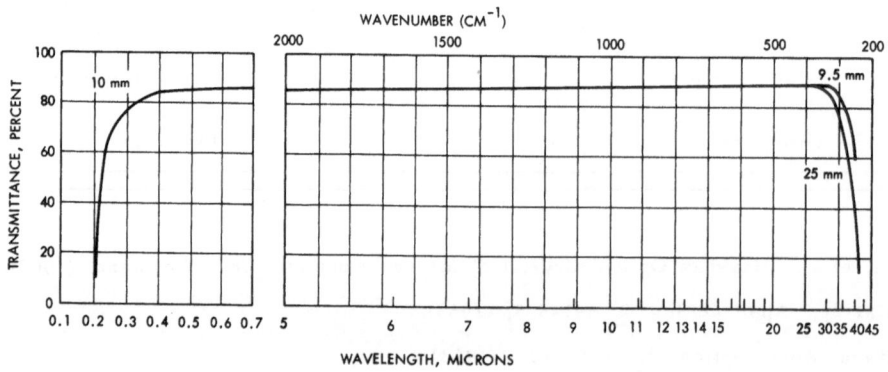

(REF. 1)

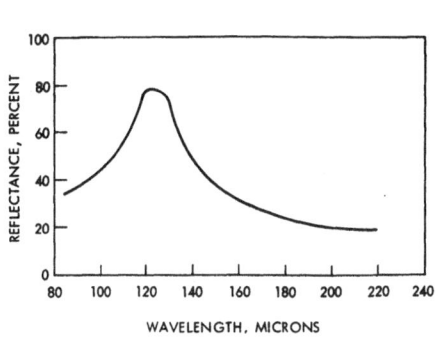

(REF. 2)

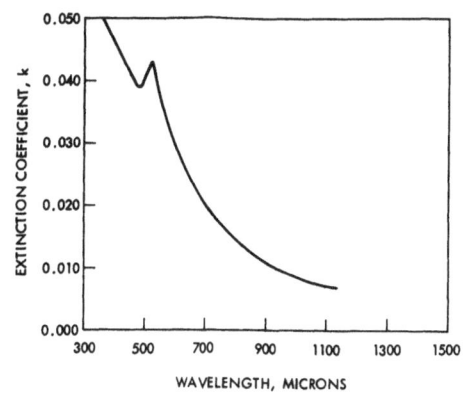

(REF. 4)

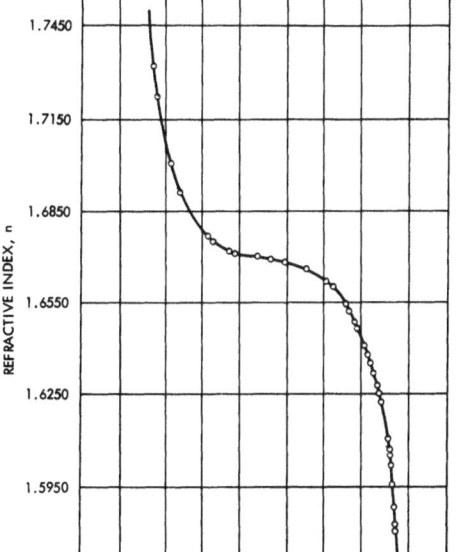

(REF. 3)

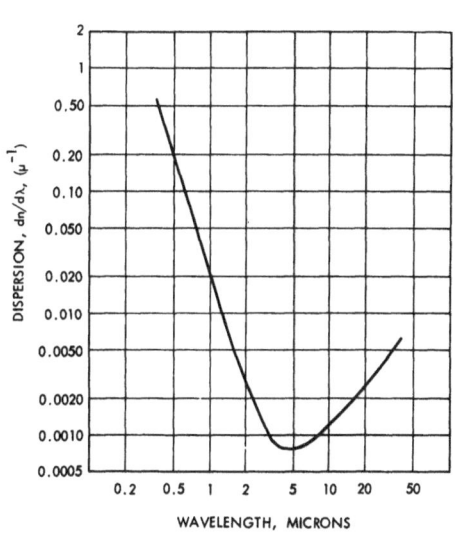

(REF. 3)

REFERENCES:

1.   A. Smakula, et al., "Harshaw Optical Crystals," Harshaw Chemical Co., Cleveland, (1967).

2.   W. M. Sinton, and W. C. Davis, J. Opt. Soc. Am., 44, 503-504, (1954).

3.   W. S. Rodney, and R. J. Spindler, J. S. Res. NBS, 51, 123-126, (1953).

4.   E. M. Dianov, Soviet Physics - Solid State, 9, 464-466, (1967).

CESIUM IODIDE

OPTICAL MATERIALS PROPERTIES
DATA SHEET

MATERIAL: CESIUM IODIDE

INTRODUCTION: This data sheet summarizes significant properties for single crystal cesium iodide.

PHYSICAL PROPERTIES, (298°K)

Density, (g/cm^3) _____ 4.51 _____

Melting/Softening
Temp. (°K) _____ 894 _____

Solubility in Water,
(g./100 g. H$_2$O) _____ 44 (273°K) _____

MECHANICAL PROPERTIES, (298°K)

Young's Modulus, (psi) _____ 0.769 x 10^6 _____

Hardness, (Knoop) _____ not available _____

THERMAL PROPERTIES, (298°K)

Linear Expansion
Coeff. (°K)$^{-1}$ _____ 50 x 10^{-6} _____

Thermal Conductivity
(10^{-4}cal/(cm sec °K) _____ 27 _____

Specific Heat, (cal/g)/°K _____ 0.048 _____

OPTICAL PROPERTIES, (298°K)

Dispersion Equation:

$$n^2 - 1 = \frac{0.34617251\lambda^2}{\lambda^2 - 0.00052701} + \frac{1.0080886\lambda^2}{\lambda^2 - 0.02149156}$$

$$+ \frac{0.28551800\lambda^2}{\lambda^2 - 0.032761} + \frac{0.39743178\lambda^2}{\lambda^2 - 0.044944}$$

$$+ \frac{3.3605359\lambda^2}{\lambda^2 - 25921}$$

Transmission Region,
(External Transmittance
≧10% with _____ 2.0 _____ mm.
thickness) _____ 0.25 - 80μ _____

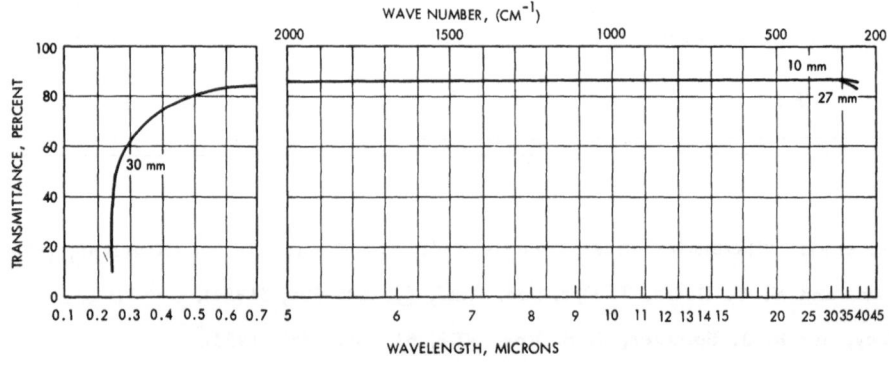

(REF. 1)

26

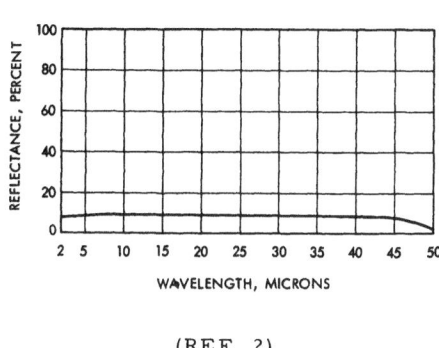

(REF. 2)

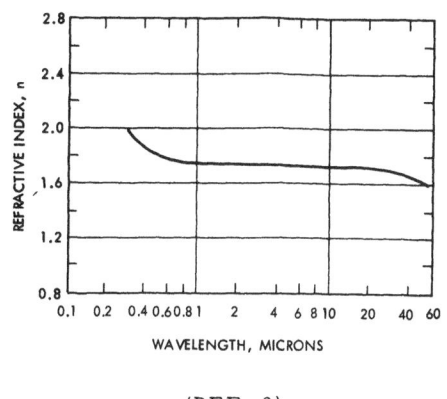

(REF. 3)

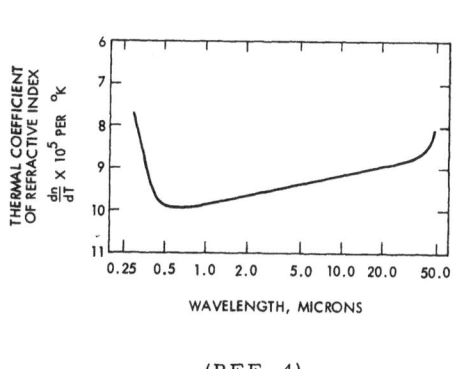

(REF. 4)

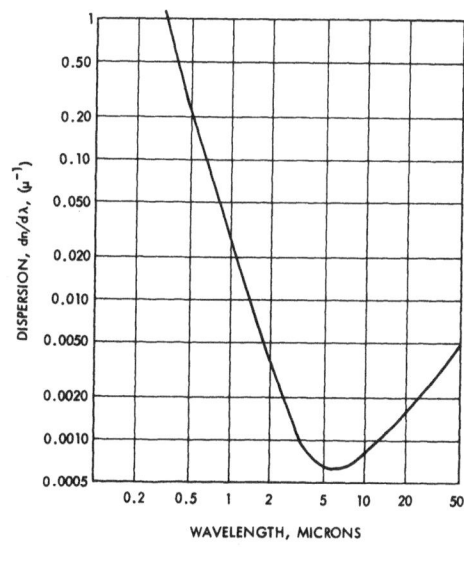

(REF. 4)

REFERENCES:

1. A. Smakula, et al., "Harshaw Optical Crystals", Harshaw Chemical Co., Cleveland, (1967).

2. D. E. McCarthy, Appl. Optics, 2, 591-595, (1963).

3. A. Smakula, Optica Acta, 9, 205-222, (1962).

4. W. S. Rodney, J. Opt. Soc. Am., 45, 987-992, (1955).

COPPER

OPTICAL MATERIALS PROPERTIES: MATERIAL: <u>COPPER</u>
DATA SHEET   (Film)

INTRODUCTION: <u>This data sheet contains optical data for copper films, in addition to thermal, physical,</u>
<u>and mechanical properties for bulk copper.</u>

PHYSICAL PROPERTIES, (298°K)

Density, (g/cm^3) _____ 8.94 _____

Melting/Softening
Temp. (°K) _____ 1356 _____

Solubility in Water,
(g./100 g. H_2O) _____ <0.01 _____

MECHANICAL PROPERTIES*, (298°K)

Young's Modulus, (psi) _____ 17 x 10^6 _____

Hardness, (Vickers) _____ 51 _____

THERMAL PROPERTIES, (298°K)

Linear Expansion
Coeff., (°K)$^{-1}$ _____ 17.7 x 10^{-6} _____

Thermal Conductivity
(10^{-4} cal/(cm sec °K) _____ 9400 _____

Specific Heat,
(cal/g)/°K _____ 0.092 _____

*Annealed bulk metal

OPTICAL PROPERTIES, (298°K)

Dispersion Equation _____ Not available _____

Transmission Region, (External
Transmittance ≥10% with _____ mm.
thickness) _____ Not available _____

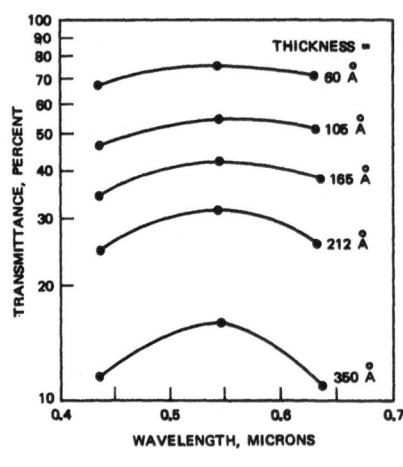

(REF. 1)

MATERIAL: <u>COPPER</u>
(Film)

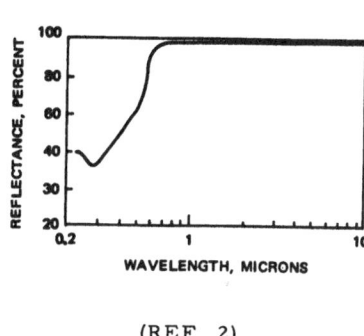

(REF. 2)

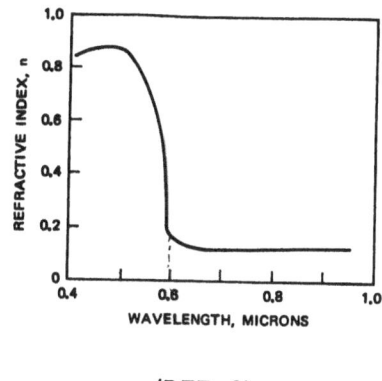

(REF. 3)

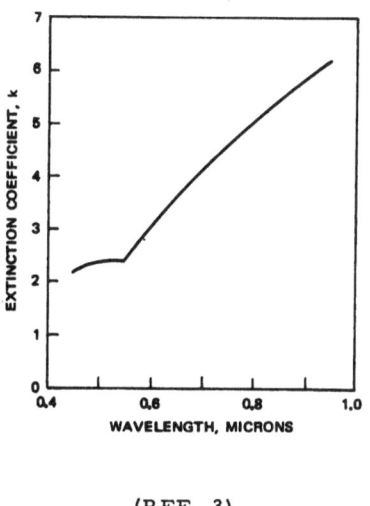

(REF. 3)

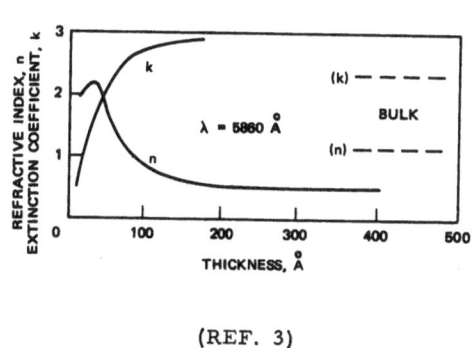

(REF. 3)

REFERENCES:

1. R.S. Adhay, Can. J. Phys. <u>38</u>, 1970-1576, (1960).

2. W.L. Wolfe, Ed., "Handbook of Military Infrared Technology," U.S. Govt. Printing Office, Washington, (1965).

3. O.S. Heavens, "Optical Properties of Thin Solid Films," Academic Press, N.Y., (1955).

CUPROUS CHLORIDE

OPTICAL MATERIALS PROPERTIES
DATA SHEET

MATERIAL: CUPROUS CHLORIDE

INTRODUCTION: This data sheet presents information on single crystal cuprous chloride.

PHYSICAL PROPERTIES, (298°K)

Density, (g/cm^3) _____ 4.14

Melting/Softening
Temp. (°K) _____ 703

Solubility in Water,
(g./100 g. H_2O) _____ 1.5

MECHANICAL PROPERTIES, (298°K)

Young's Modulus, (psi) _____ Not available

Hardness (microhardness) _____ 11 kg/mm^2

THERMAL PROPERTIES, (298°K)

Linear Expansion
Coeff., (°K)$^{-1}$ _____ 13.6 x 10^{-6}

Thermal Conductivity
(10^{-4} cal/(cm sec °K) _____ Not available

Specific Heat,
(cal/g)/°K _____ 0.12

OPTICAL PROPERTIES, (298°K)

Dispersion Equation _____ Not available

Transmission Region, (External
Transmittance ≥10% with _9.1_ mm.
thickness) _____ 0.4 - 19µ

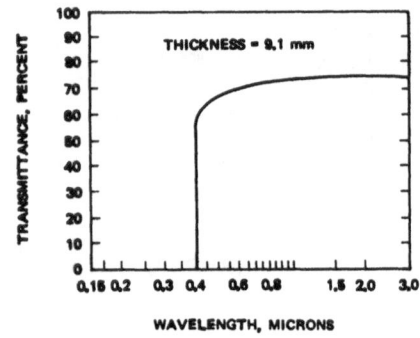

(REF. 1)

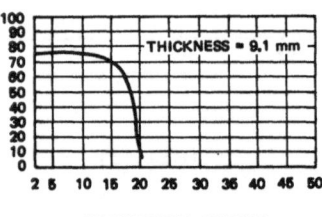

(REF. 2)

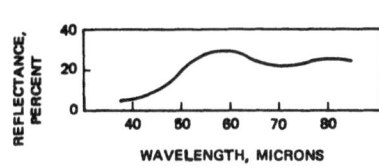

(REF. 3)

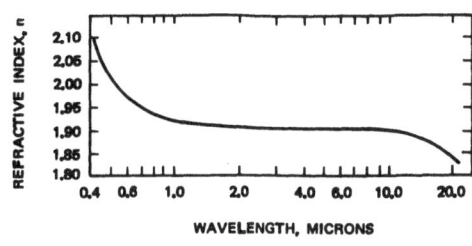

(REF. 4)

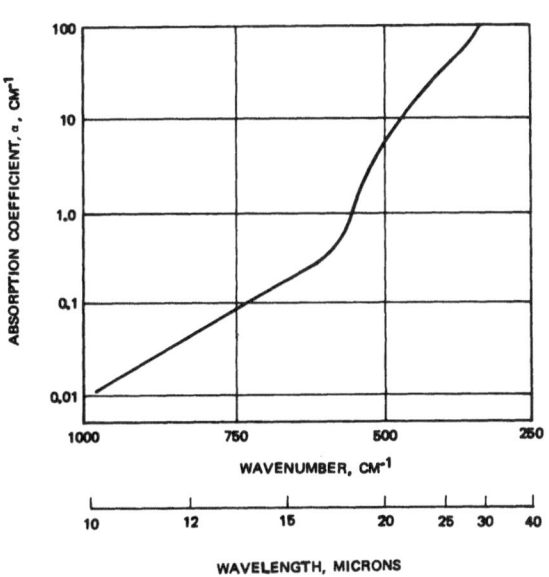

(REF. 5)

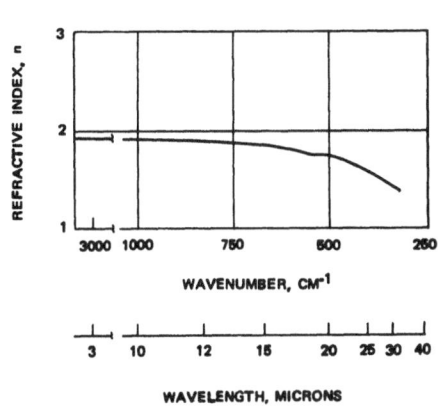

(REF. 5)

REFERENCES:

1. D.E. McCarthy, Appl. Optics, 6, 1896-1898, (1967).

2. D.E. McCarthy, Appl. Optics, 4, 317-320, (1965).

3. J.N. Plendl, et al, Appl. Optics, 5, 397-401, (1966).

4. A. Feldman and D. Horowitz, Opt. Soc. Am., 59, 1406-1408, (1969).

5. A. Smakula, Report No. AD 663 745, (1967).

GALLIUM ARSENIDE

OPTICAL MATERIALS PROPERTIES
DATA SHEET

MATERIAL: GALLIUM
ARSENIDE

INTRODUCTION: This data sheet contains information for intrinsic single crystal gallium arsenide.

PHYSICAL PROPERTIES, (298°K)

Density, (g/cm^3)_____5.307_____

Melting/Softening
Temp. (°K)_____1511_____

Solubility in Water,
(g./100 g. H$_2$O)_____<0.005_____

MECHANICAL PROPERTIES, (298°K)

Young's Modulus, (psi) _not available_

Hardness, (Knoop)_____750_____

THERMAL PROPERTIES, (298°K)

Linear Expansion
Coeff. (°K)$^{-1}$_____6.8 x 10^{-6}_____

Thermal Conductivity
(10^{-4} cal/(cm sec °K)_1080_

Specific Heat,
(cal/g)/°K _____0.076_____

OPTICAL PROPERTIES, (298°K)

Dispersion Equation:_not available_

Transmission Region, (External
Transmittance ≥ 10% with___2.0___mm.
thickness)_____1.0 - 15μ_____

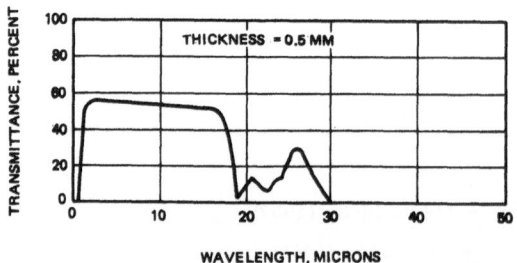

(REF. 1)

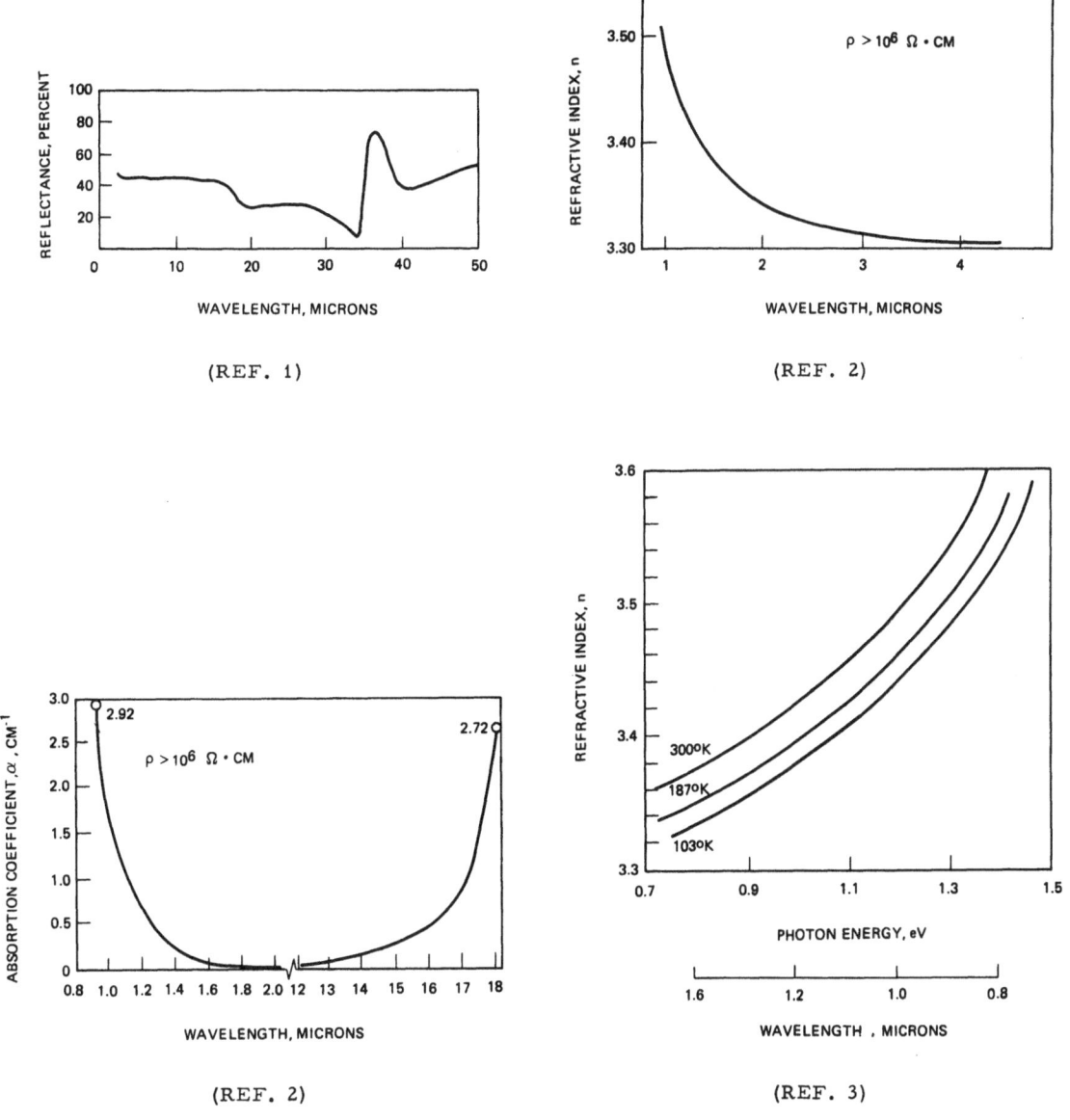

(REF. 1)

(REF. 2)

(REF. 2)

(REF. 3)

REFERENCES:

1. D.E. McCarthy, Appl. Optics, 7, 1997-2000, (1968).

2. E. Hirschmann, and T.E. Walsh, Report No. NASA TN D-4049, (June 1967).

3. D.T.F. Marple, J. Appl. Phys. 35, 1241-1242, (1964).

GERMANIUM

OPTICAL MATERIALS PROPERTIES
DATA SHEET

MATERIAL: <u>GERMANIUM</u>

INTRODUCTION: <u>This data sheet summarizes results for intrinsic single crystal germanium.</u>

PHYSICAL PROPERTIES, (298°K)

Density, (g/cm^3) _____ 5.33

Melting/Softening
Temp. (°K) _____ 1209

Solubility in Water,
(g./100 g. H$_2$O) _____ <0.005

MECHANICAL PROPERTIES, (298°K)

Young's Modulus, (psi) _14.9 x 10^6_

Harness, (Knoop) _____ 700-880

THERMAL PROPERTIES, (298°K)

Linear Expansion
Coeff. (°K)$^{-1}$ _____ 5.5 x 10-6

Thermal Conductivity
(10^{-4} cal/(cm sec °K) _____ 1400

Specific Heat,
(cal/g)/°K _____ 0.074

OPTICAL PROPERTIES, (298°K)

Dispersion Equation

$$n = A + BL + CL^2 + D\lambda^2 + E\lambda^2$$

where

A = 3.99931; B = 0.391707; C = 0.163492;
D = -0.0000060; E = 0.000000053
(valid between 2.0 and 13.5μ)

Transmission Region, (External
Transmittance ≥10% with _2.0_ mm.
thickness) _1.8 - 23μ_

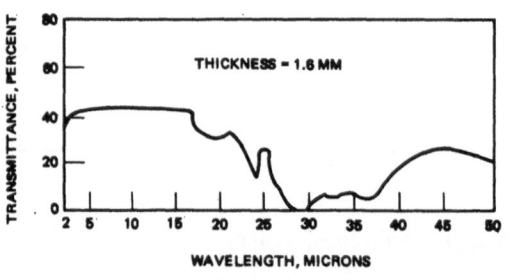

(REF. 1)

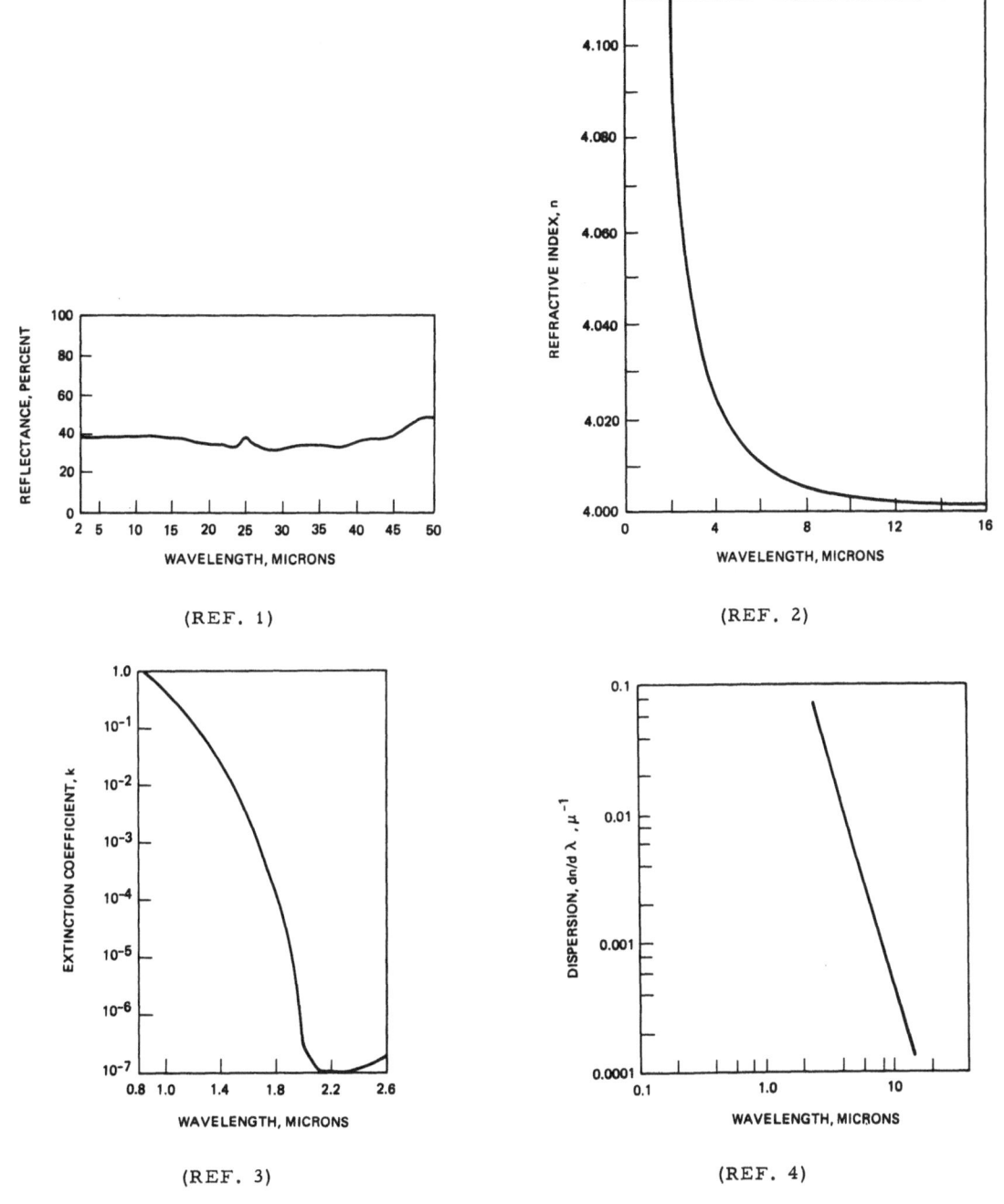

(REF. 1)

(REF. 2)

(REF. 3)

(REF. 4)

REFERENCES:

1. D. E. McCarthy, Appl. Optics, <u>2</u>, 591-595, (1963).

2. C. D. Salzberg and J. J. Villa, J. Opt. Soc. Am., <u>47</u>, 244-246, (1957).

3. H. B. Briggs, J. Opt. Soc. Am., <u>42</u>, 686-687, (1952).

4. W. L. Wolfe, Ed., "Handbook of Military Infrared Technology", U. S. Govt. Printing Office, Washington, (1965).

GOLD

OPTICAL MATERIALS PROPERTIES   MATERIAL: <u>GOLD</u>
DATA SHEET   (FILM)

INTRODUCTION: <u>This data sheet presents information for gold film, formed by evaporation. The</u>
<u>optical data may be compromised by uncertainty concerning the crystallinity of the film.</u>

PHYSICAL PROPERTIES, (298°K)

Density, (g/cm^3)_____19.32

Melting/Softening
Temp. (°K)_____1336

Solubility in Water,
(g./100 g. H$_2$O)_____<0.001

MECHANICAL PROPERTIES, (298°K)

Young's Modulus, (psi)___0.78 x 10^6

Hardness, (Vickers)_____50-120

THERMAL PROPERTIES, (298°K)

Linear Expansion
Coeff. (°K)$^{-1}$_____14.1 x 10^{-6}

Thermal Conductivity
(10^{-4}ca/(cm sec °K)___7100

Specific Heat,
(cal/g)/°K _____0.031

OPTICAL PROPERTIES, (298°K)

Dispersion Equation:_not available_

Transmission Region, (External
Transmittance ≥ 10% with_____mm.
thickness)_____not available

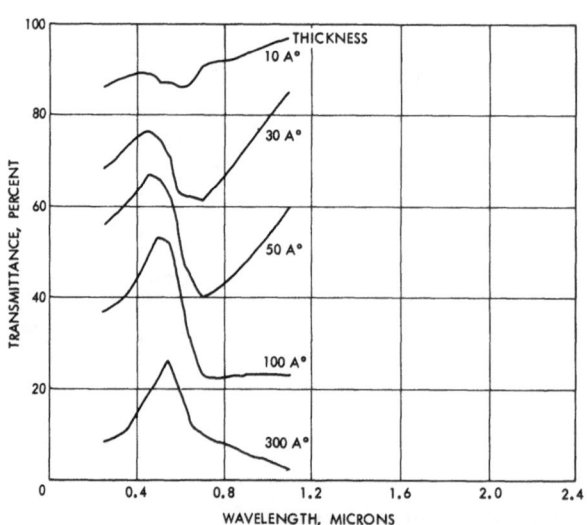

(REF. 1)

36

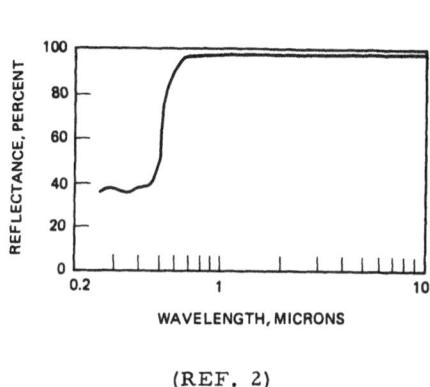

(REF. 2)

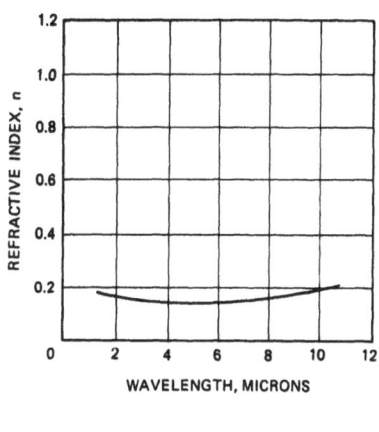

(REF. 3)

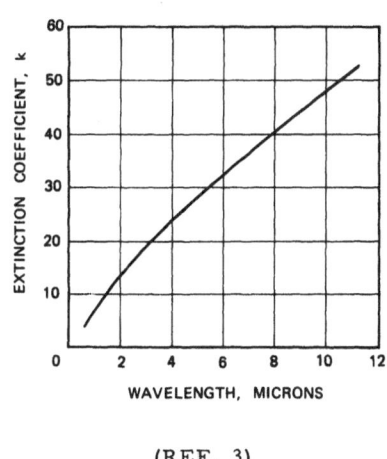

(REF. 3)

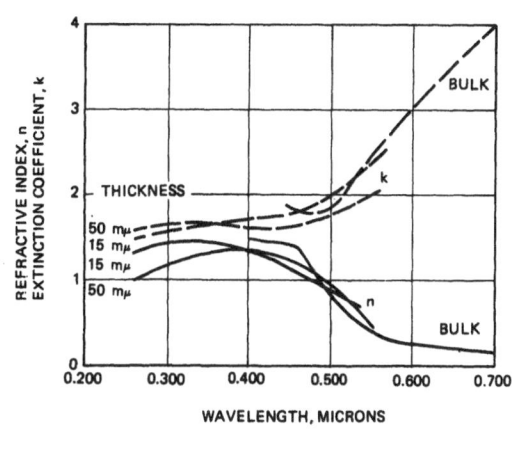

(REF. 3)

REFERENCES:

1.   F. Goos, Z. Phys., 106, 606-619, (1937).

2.   W.L. Wolfe, Ed., "Handbook of Military Infrared Technology", U.S. Govt. Printing Office, Washington, (1965).

3.   O.S. Heavens, in "Reports on Progress in Physics", A.C. Sickland, Ed., 23, 30-65, (1960).

INDIUM ARSENIDE

OPTICAL MATERIALS PROPERTIES
DATA SHEET

MATERIAL: <u>INDIUM</u>
<u>ARSENIDE</u>

INTRODUCTION: This data sheet contains information for single crystal indium arsenide.

PHYSICAL PROPERTIES, (298°K)

Density, (g/cm^3) 5.66

Melting/Softening
Temp. (°K) 1215

Solubility in Water,
(g./100 g. H$_2$O <0.05

MECHANICAL PROPERTIES, (298°K)

Young's Modulus, (psi) Not available

Hardness, (Knoop) 380

THERMAL PROPERTIES, (298°K)

Linear Expansion
Coeff., (°K)$^{-1}$ 5.3 x 10^{-6}

Thermal Conductivity
(10^{-4} cal/(cm sec °K) 640 (polycrystalline)

Specific Heat,
(cal/g)/°K 0.061

OPTICAL PROPERTIES, (298°K)

Dispersion Equation Not available

Transmission Region, (External
Transmittance ≥10% with 2.0 mm.
thickness) 3.8 - 7.0μ

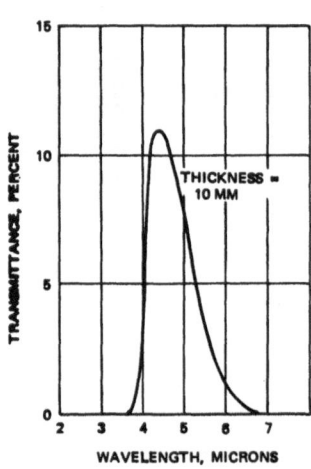

(REF. 1)

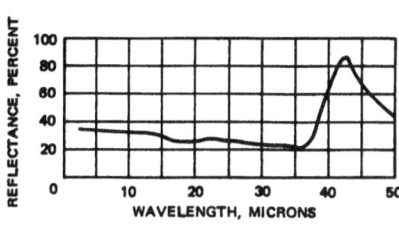

(REF. 2)

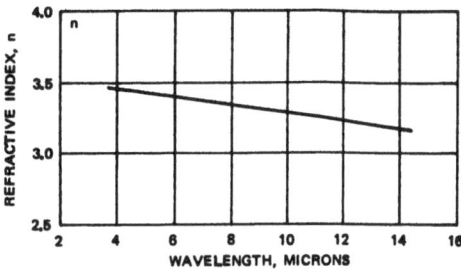

(REF. 1)

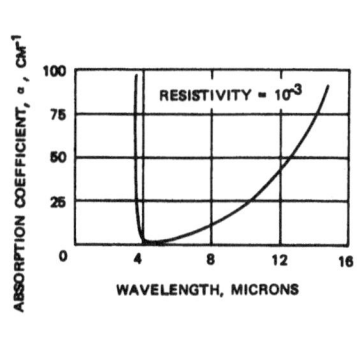

(REF. 3)

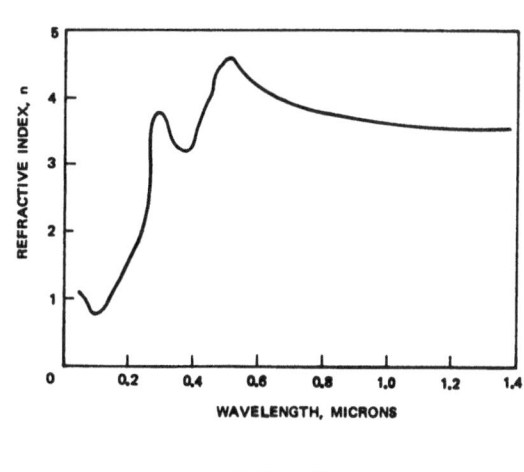

(REF. 4)

REFERENCES:

1. P. Billard, Acta Electronica, 6, 75-169, (1962).

2. D.E. McCarthy, Appl. Optics, 7, 1997-2000, (1968).

3. F. Oswald and R. Schade, Z. Naturforsch, 9a, 611-617, (1954).

4. B.O. Seraphin and H.E. Bennett in "Semiconductors and Semimetals," R. K. Willard and R. C. Beer, Eds, Vol. 3, Academic Press, N.Y., (1967).

LEAD SULFIDE

OPTICAL MATERIALS PROPERTIES
DATA SHEET

MATERIAL: LEAD
SULFIDE

INTRODUCTION: This data sheet furnishes information on lead sulfide with optical data applying to single crystal lead sulfide.

PHYSICAL PROPERTIES, (298°K)

Density, (g/cm^3) ___7.5___

Melting/Softening
Temp. (°K)___1392___

Solubility in Water,
$(g./100 g. H_2O)$ ___3×10^{-6}___

MECHANICAL PROPERTIES, (298°K)

Young's Modulus, (psi)___1.64×10^{-6}___

Hardness, (Mohs)___2.5___

THERMAL PROPERTIES, (298°K)

Linear Expansion
Coeff. $(°K)^{-1}$ ___24×10^{-6}___

Thermal Conductivity
$(10^{-4} cal/(cm \ sec \ °K)$ ___500___

Specific Heat,
$(cal/g)/°K$ ___0.039___

OPTICAL PROPERTIES, (298°K)

Dispersion Equation:___not available___

Transmission Region, (External
Transmittance <10% with ___2.0___ mm.
thickness) ___3.0 - 7.0µ___

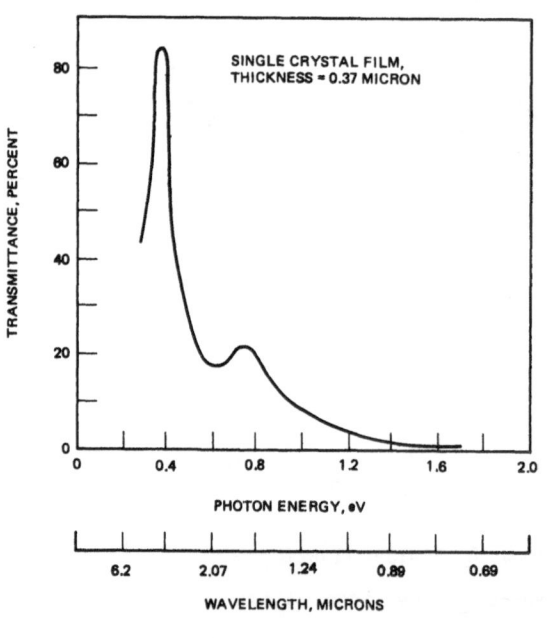

(REF. 1)

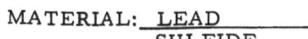

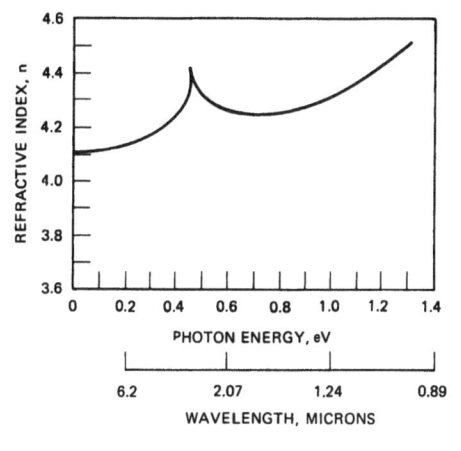

(REF. 1)

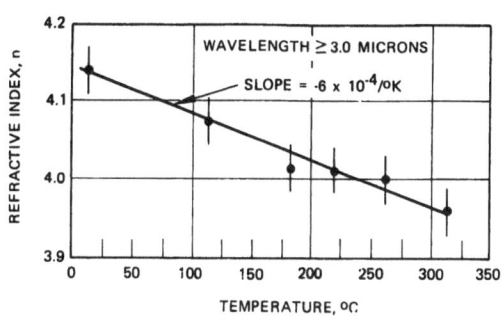

(REF. 3)

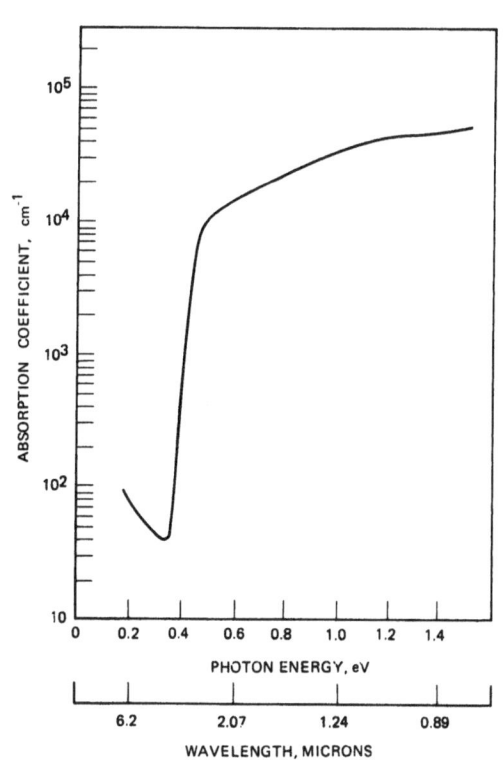

(REF. 1)

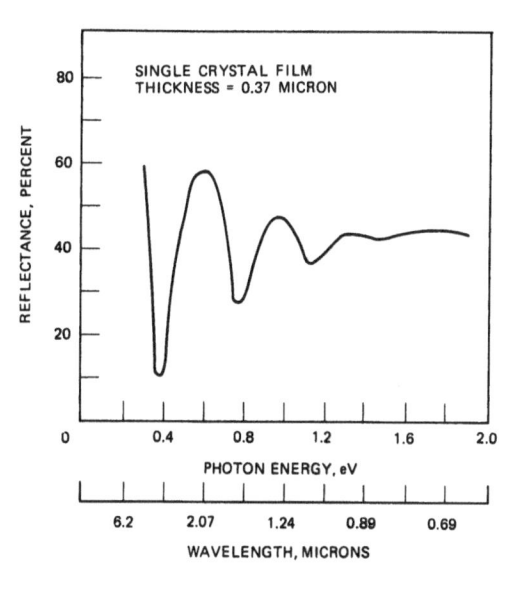

(REF. 1)

REFERENCES:

1. R.B. Schoolar and J.R. Dixon, Phys. Rev., <u>137</u>, A667-A670, (1965).

2. W.W. Scanlon, Phys. Rev., <u>109</u>, 47-50, (1958).

3. D.G. Avery, Proc. Phys. Soc. (London), <u>B67</u>, 2-8, (1954).

LITHIUM FLUORIDE

OPTICAL MATERIALS PROPERTIES
DATA SHEET

MATERIAL: <u>LITHIUM</u>
<u>FLUORIDE</u>

INTRODUCTION: <u>This data sheet presents information for single crystal lithium fluoride.</u>

PHYSICAL PROPERTIES, (298°K)

Density, (g/cm³)____<u>2.64</u>____

Melting/Softening
Temp. (°K)____<u>1403</u>____

Solubility in Water,
(g./100 g. H₂O)____<u>0.27 (291°K)</u>____

MECHANICAL PROPERTIES, (298°K)

Young's Modulus, (psi)__<u>9.4 x 10⁶</u>__

Hardness, (Knoop)____<u>~113 (600 g.)</u>____

THERMAL PROPERTIES, (298°K)

Linear Expansion
Coeff. (°K)⁻¹____<u>37 x 10⁻⁶</u>____

Thermal Conductivity
(10⁻⁴ cal/(cm sec °K)__<u>340</u>__

Specific Heat,
(cal/g)/°K____<u>0.373 (283°K)</u>____

OPTICAL PROPERTIES, (298°K)

Dispersion Equation:

$$n = A + BL + CL^2 + D\lambda^2 + E\lambda^4$$

where

A = 1.38761; B = 0.001796; C = -0.000041;
D = -0.0023045; E = -0.00000557
(between 0.5 and 6.0μ).

Transmission Region, (External
Transmittance ≥ 10% with __<u>2.0</u>__ mm.
thickness)____<u>0.12 - 9.0μ</u>____

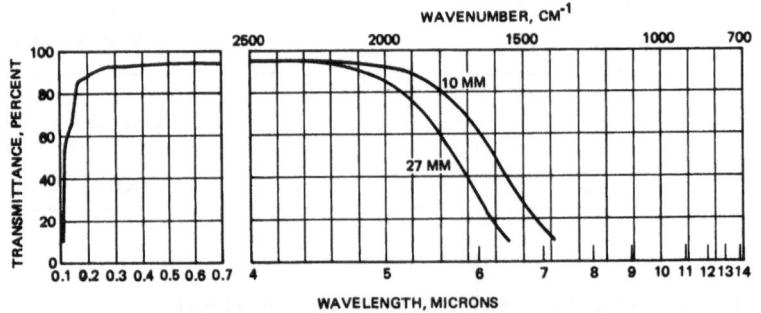

(REF. 1)

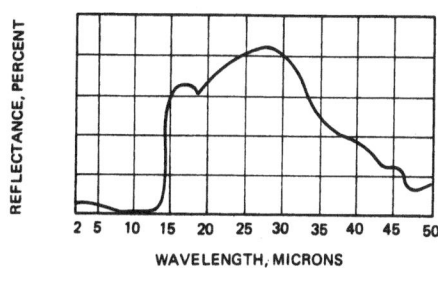

(REF. 2)

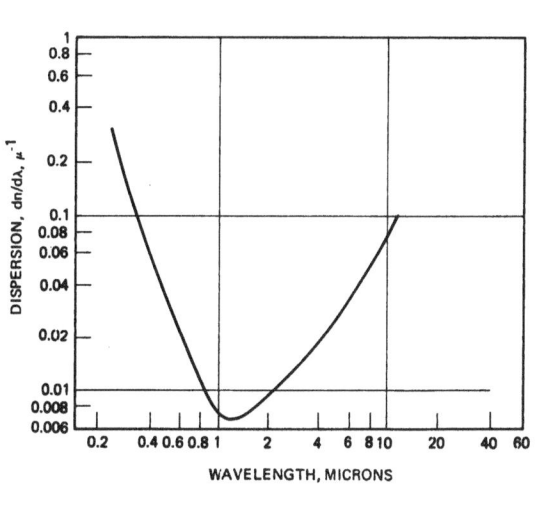

(REF. 3)

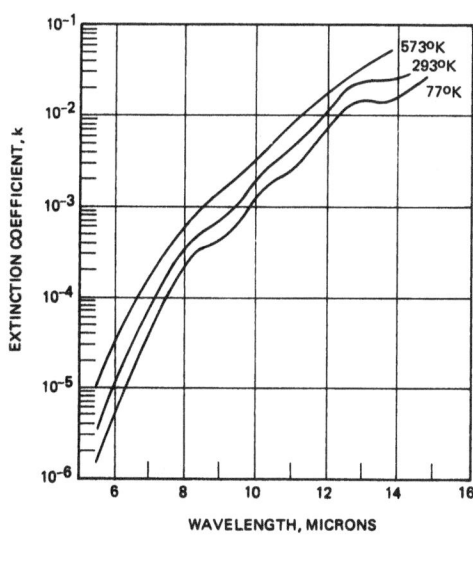

(REF. 4)

(REF. 3)

REFERENCES:

1. A. Smakula, "Harshaw Optical Crystals", The Harshaw Chemical Co., Cleveland, (1967).

2. D.E. McCarthy, Appl. Optics, 2, 591-595, (1963).

3. A. Smakula, Optica Acta, 9, 205-222, (1962).

4. M. Klier, Z. Physik, 150, 49-63, (1958).

LITHIUM NIOBATE

OPTICAL MATERIALS PROPERTIES
DATA SHEET

MATERIAL: LITHIUM
NIOBATE

INTRODUCTION: This data sheet contains information for single crystal lithium niobate.

PHYSICAL PROPERTIES, (298°K)

Density, (g/cm^3)___4.70___

Melting/Softening
Temp. (°K)___1533___

Solubility in Water,
(g./100 g. H$_2$O)___<0.005___

MECHANICAL PROPERTIES, (298°K)

Young's Modulus, (psi)___not available___

Hardness, (Mohs) ___~5___

THERMAL PROPERTIES, (298°K)

Linear Expansion
Coeff. (°K)$^{-1}$___16.7 x 10^{-6}___

Thermal Conductivity
(10^{-4} cal/(cm sec °K)___100___

Specific Heat,
(cal/g)/°K ___0.153___

OPTICAL PROPERTIES, (298°K)

Dispersion Equation: not available

Transmission Region, (External
Transmittance ≥10% with___10___mm.
thickness)___0.33 - 5.2µ___

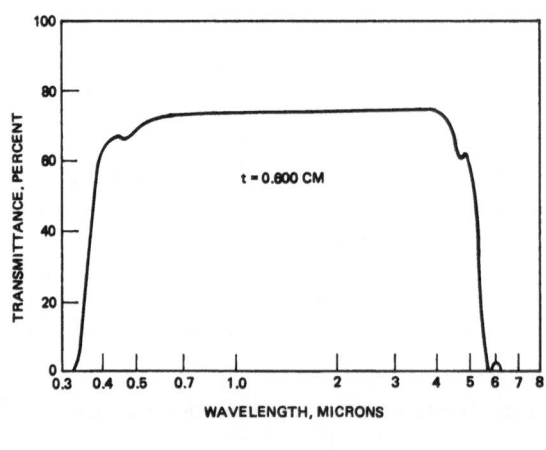

(REF. 1)

44

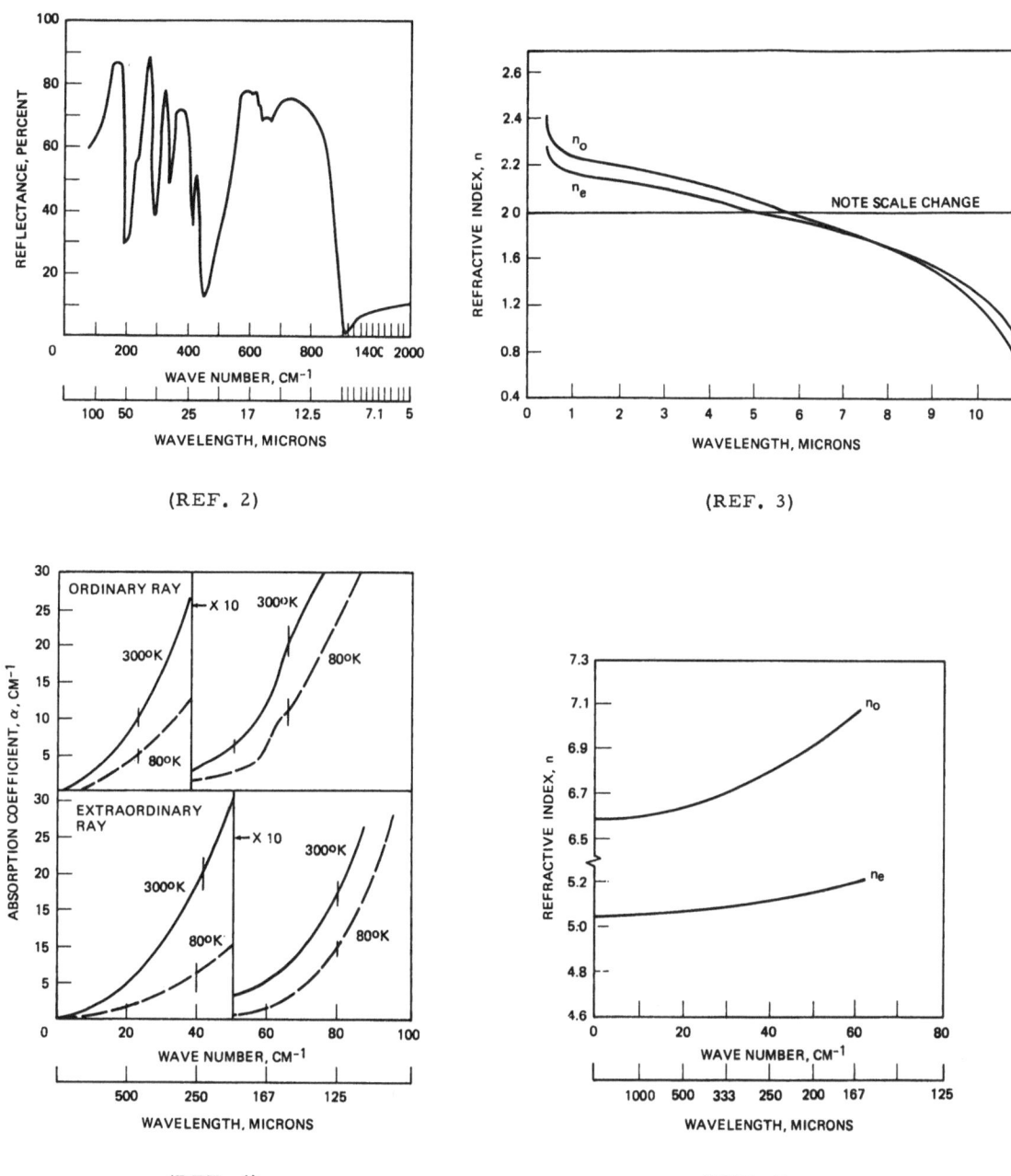

(REF. 2)

(REF. 3)

(REF. 4)

(REF. 4)

REFERENCES:

1. K. Nassau, et al., J. Phys. Chem. Solids, <u>27</u>, 989-996, (1966).

2. J.D. Axe and D.F. O'Kane, Appl. Phys. Letters, <u>9</u>, 58-60, (1966).

3. A.S. Barker, Jr. and R. Loudon, Phys. Rev. <u>158</u>, 433-445, (1967).

4. D.R. Bosomworth, Appl. Phys. Letters, <u>9</u>, 330-331, (1966).

MAGNESIUM FLUORIDE
(Film)

<table>
<tr><td colspan="2">OPTICAL MATERIALS PROPERTIES
DATA SHEET</td><td>MATERIAL: <u>MAGNESIUM</u>
<u>FLUORIDE</u>
<u>(Film)</u></td></tr>
</table>

INTRODUCTION: This data sheet applies to freshly evaporated films of magnesium fluoride deposited on glass.

PHYSICAL PROPERTIES, (298°K)

Density, (g/cm^3)_____~3.2_____

Melting/Softening
Temp. (°K)_____1528_____

Solubility in Water,
(g./100 g. H$_2$O)_____Not available_____

MECHANICAL PROPERTIES, (298°K)

Young's Modulus, (psi)____Not available____

Hardness, Knoop)_____Not available_____

THERMAL PROPERTIES, (298°K)

Linear Expansion
Coeff., (°K)$^{-1}$_____Not available_____

Thermal Conductivity
(10^{-4} cal/(cm sec °K)____Not available____

Specific Heat,
(cal/g)/°K_____Not available_____

OPTICAL PROPERTIES, (298°K)

Dispersion Equation_____Not available_____

Transmission Region, (External
Transmittance ≥10% with __~1/4 λ__
thickness)_____0.2 - 5.0μ_____

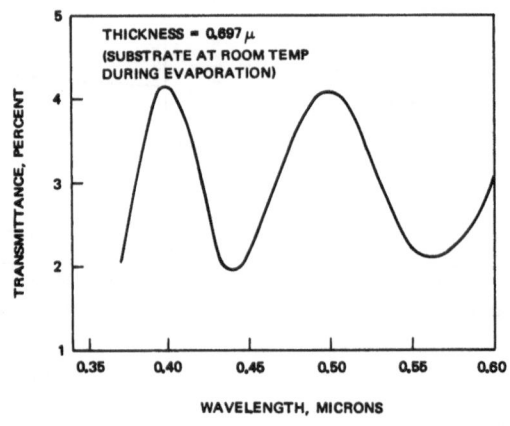

THICKNESS = 0.697 μ
(SUBSTRATE AT ROOM TEMP
DURING EVAPORATION)

TRANSMITTANCE, PERCENT

WAVELENGTH, MICRONS

(REF. 1)

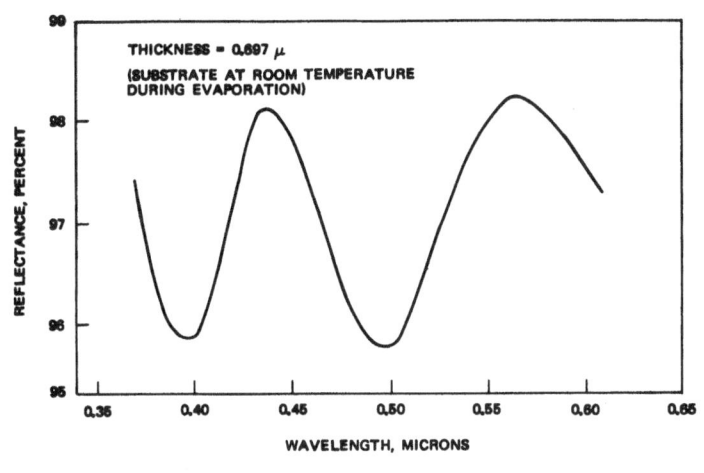

(REF. 1)

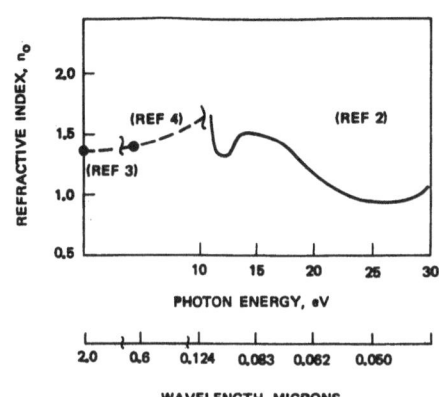

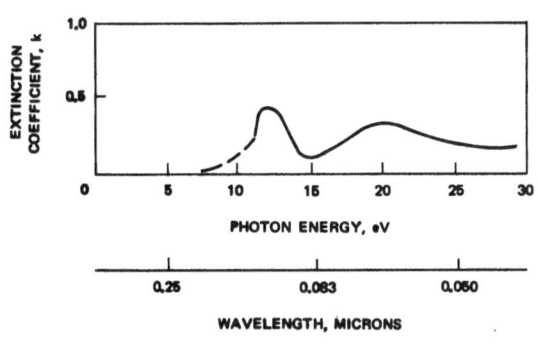

(REF. 2)

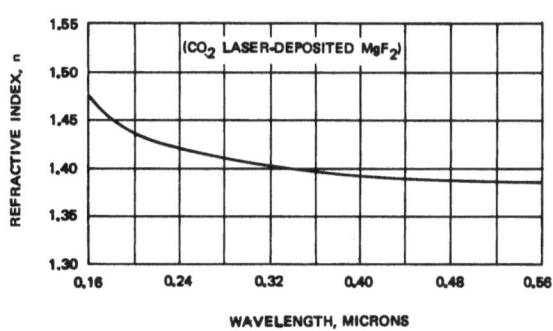

(REF. 5)

REFERENCES:

1. N. Morita, J. Phys. Soc., Japan, <u>11</u>, 975-980, (1956).

2. M.W. Williams, et al, J. Appl. Phys., <u>38</u>, 1701-1705, (1967).

3. J.R. Jenness, Jr., J. Am. Opt. Soc., <u>46</u>, 157-159, (1956).

4. J.F. Hall, Jr., and W.F.C. Ferguson, J. Am. Opt. Soc., <u>45</u>, 74-75, (1955).

5. G. Hass and J.B. Ramsey, Appl. Optics, <u>8</u>, 1115-1118, (1969).

MAGNESIUM FLUORIDE
(Single Crystal)

OPTICAL MATERIALS PROPERTIES
DATA SHEET

MATERIAL: <u>MAGNESIUM</u>
<u>FLUORIDE</u>
(Single crystal)

INTRODUCTION: <u>This data sheet contains information for single crystal magnesium fluoride</u>

PHYSICAL PROPERTIES, (298°K)

Density, (g/cm^3) _____ 3.177 _____

Melting/Softening
Temp. (°K) _____ 1498 _____

Solubility in Water,
(g./100 g. H_2O) _____ 0.0076 (291°K) _____

MECHANICAL PROPERTIES, (298°K)

Young's Modulus, (psi) __not available__

Hardness, (Mohs) _____ 415 (100 g.) _____

THERMAL PROPERTIES, (298°K)

Linear Expansion
Coeff. (°K)$^{-1}$ _____ 4×10^{-6} _____

Thermal Conductivity
(10^{-4} cal/(cm sec °K) __not available__

Specific Heat,
(cal/g)/°K _____ 0.284 _____

OPTICAL PROPERTIES, (298°K)

Dispersion Equation:

$n_o = 1.36957 + 0.0035821/[\lambda - 0.14925]$

$n_e = 1.38100 + 0.0037415/[\lambda - 0.14947]$

where

λ = wavelength between 0.4 and 0.7 microns

Transmission Region, (External
Transmittance ≥10% with _____ 2.0 _____ mm.
thickness) _____ 0.1 - 9.7μ _____

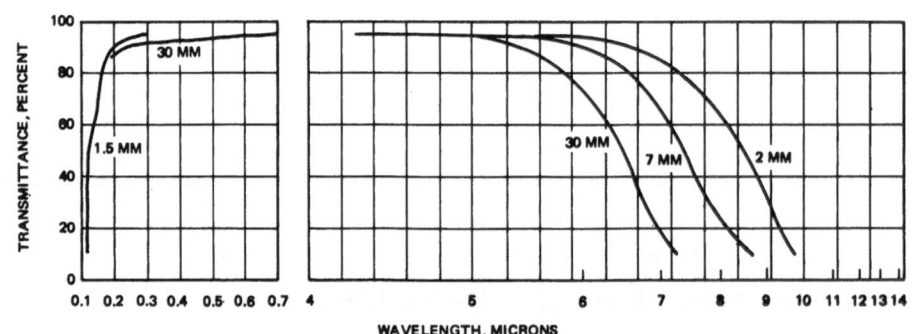

(REF. 1)

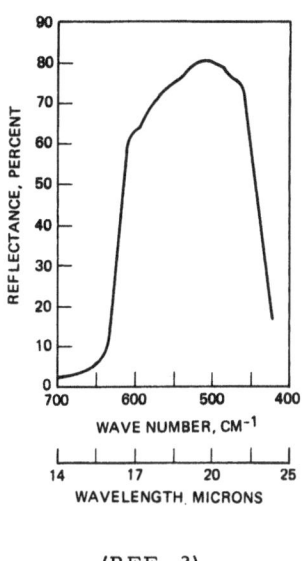

(REF. 2)

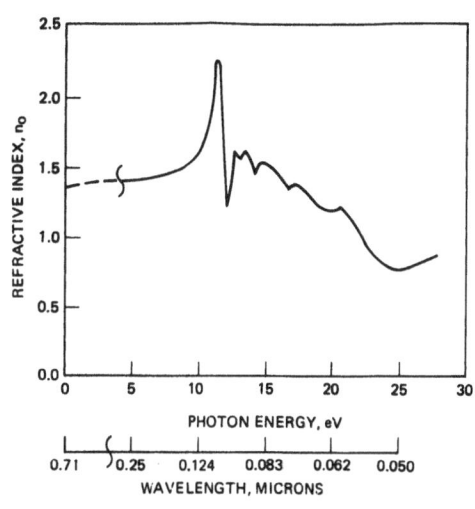

(REF. 2, 3)

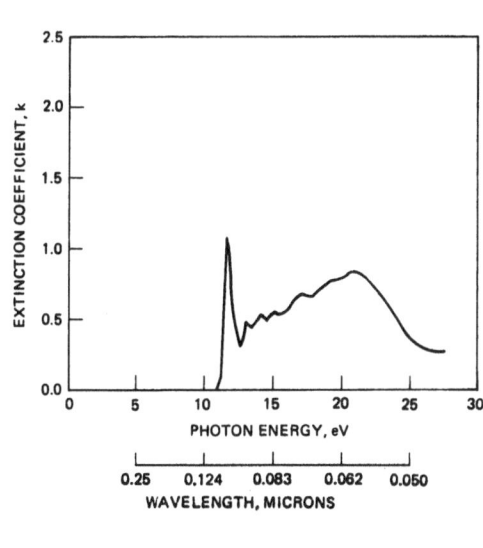

(REF.)

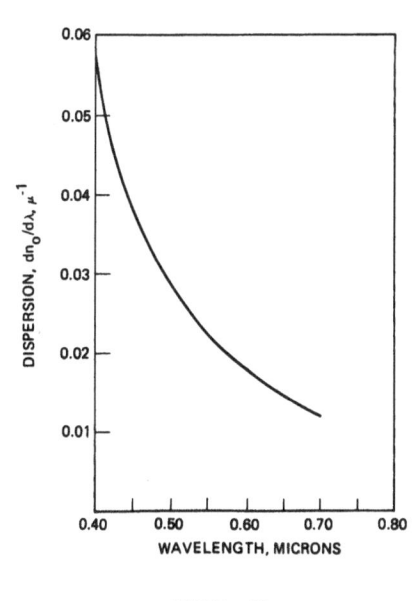

(REF. 2)

REFERENCES:

1. A. Smakula, et al., "Harshaw Optical Crystals", The Harshaw Chemical Co., Cleveland, (1967).

2. A. Duncanson and R.W. Stevenson, Proc. Royal Soc. London, 72, 1001-1006, (1958).

3. M.W. Williams, et al., J.Appl. Phys., 38, 1701-1705, (1967).

MAGNESIUM OXIDE

OPTICAL MATERIALS PROPERTIES
DATA SHEET

MATERIAL: <u>MAGNESIUM
OXIDE</u>

INTRODUCTION: <u>This data sheet contains information for single crystal magnesium oxide.</u>

PHYSICAL PROPERTIES, (298°K)

Density, (g/cm^3) _____ 3.58 _____

Melting/Softening
Temp. (°K) _____ 3073 _____

Solubility in Water,
(g./100 g. H$_2$O) _____ 1.2 x 10^{-5} (293°K) _____

MECHANICAL PROPERTIES, (298°K)

Young's Modulus, (psi) _48.2 x 10^6_

Hardness, (Knoop) _____ 692 (600 g.) _____

THERMAL PROPERTIES, (298°K)

Linear Expansion
Coeff. (°K)$^{-1}$ _____ 10.5 x 10^{-6} _____

Thermal Conductivity
(10^{-4} cal/(cm sec °K) _600_

Specific Heat,
(cal/g)/°K _____ 0.222 _____

OPTICAL PROPERTIES, (298°K)

Dispersion Equation:

$$n^2 = 2.956362 - 0.1062387\lambda^2 - 0.0000204968\lambda^4 - \frac{0.02195770}{\lambda^2 - 0.01428322}$$

Transmission Region, (External
Transmittance ≥ 10% with _2.0_ mm.
thickness _0.25 - 8.5μ_

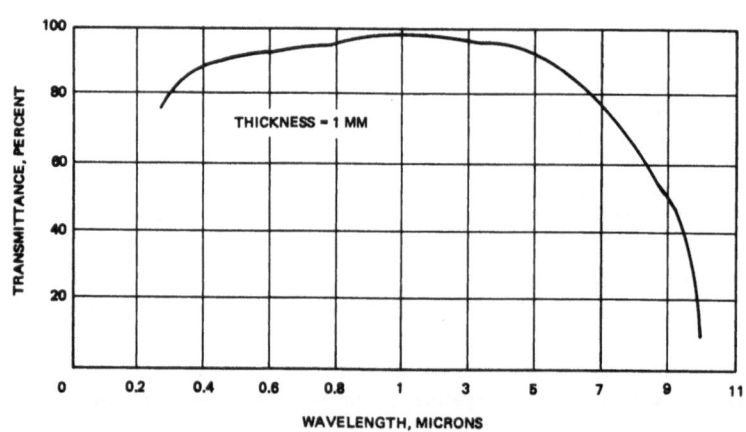

THICKNESS = 1 MM

(REF. 1)

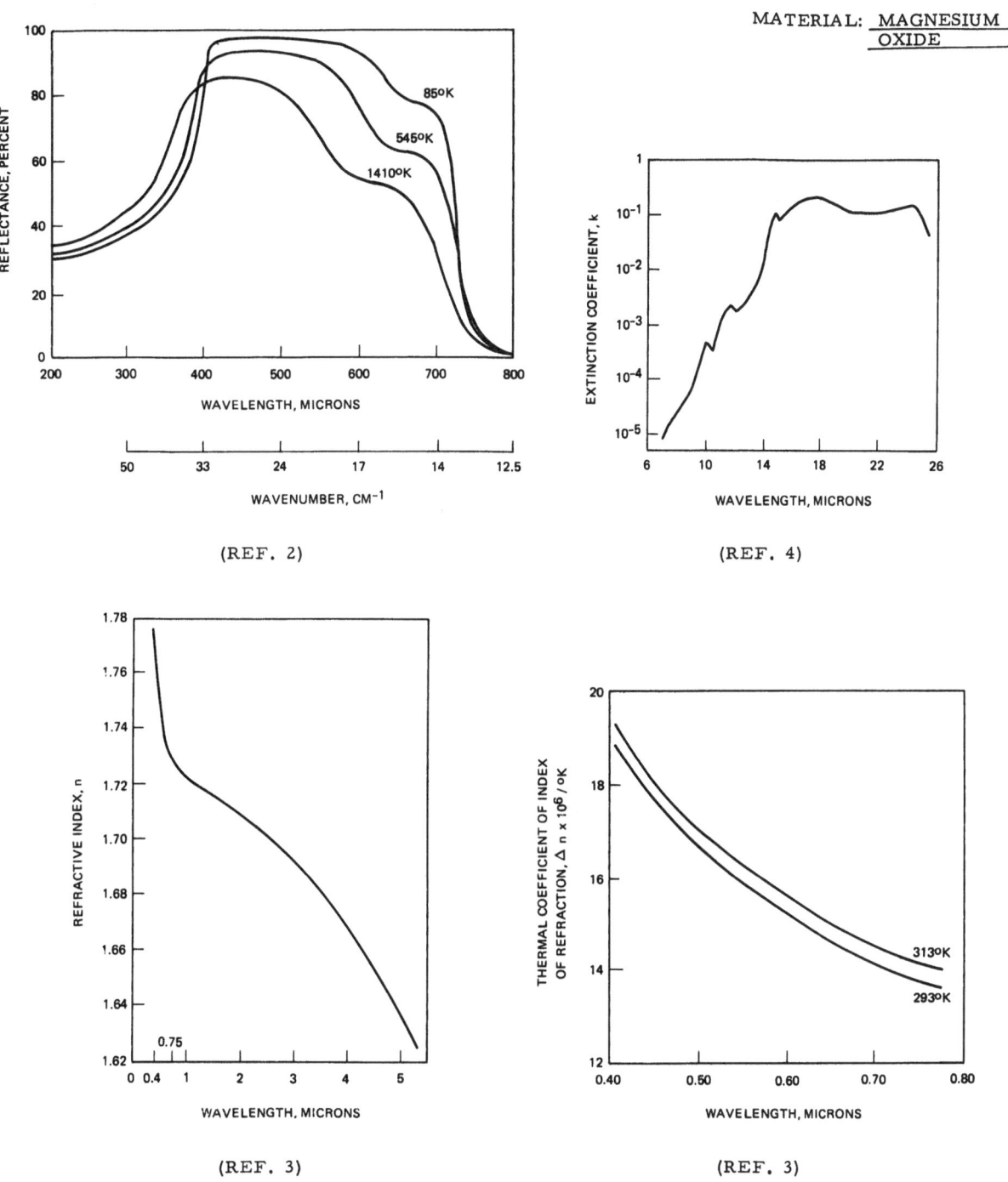

(REF. 2)

(REF. 4)

(REF. 3)

(REF. 3)

REFERENCES:

1. Data Sheet, Semi-Elements, Inc., Saxonburg, Pa.

2. A. Kahan, Report No. AFCRL-66-327, (1966).

3. R.E. Stephens and I.H. Malitson, J. Res. NBS, 49, 249-252, (1952).

4. J.C. Wilmott, Proc. Phys. Soc. London, 63A, 389-402, (1950).

PALLADIUM

OPTICAL MATERIALS PROPERTIES
DATA SHEET

MATERIAL: <u>PALLADIUM</u>
<u>(Film)</u>

INTRODUCTION: This data sheet contains optical data for palladium films as well as thermal, physical and mechanical properties for bulk palladium.

PHYSICAL PROPERTIES, (298°K)

Density, (g/cm^3) ___ 12.02 ___

Melting/Softening
Temp. (°K) ___ 1827 ___

Solubility in Water,
(g./100 g. H$_2$O) ___ <0.01 ___

MECHANICAL PROPERTIES*, (298°K)

Young's Modulus, (psi) ___ 1.6 x 10^7 ___

Hardness, (Vickers) ___ 37 ___

THERMAL PROPERTIES, (298°K)

Linear Expansion
Coeff., (°K)$^{-1}$ ___ 11.67 x 10^{-6} ___

Thermal Conductivity
(10^{-4} cal/(cm sec °K) ___ 1680 ___

Specific Heat,
(cal/g)/°K ___ 0.0584 ___

*Annealed bulk metal

OPTICAL PROPERTIES, (298°K)

Dispersion Equation ___ Not available ___

Transmission Region, (External
Transmittance ≥10% with ___ mm.
thickness) ___ Not available ___

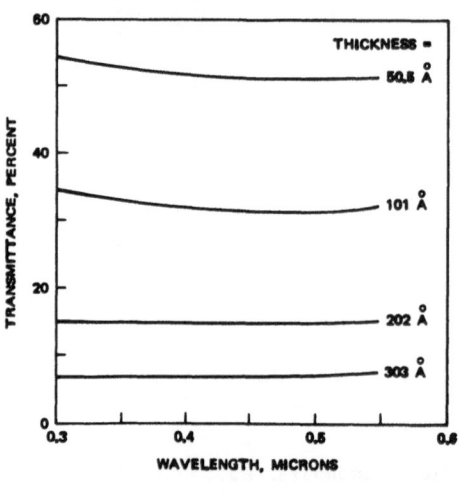

(REF. 1)

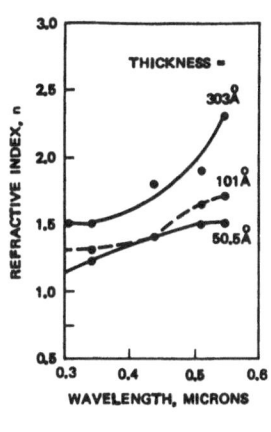

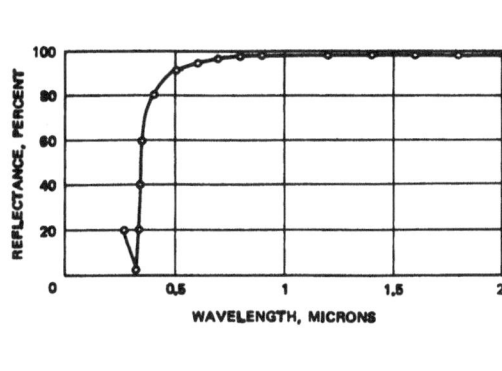

(REF. 2)

(REF. 1)

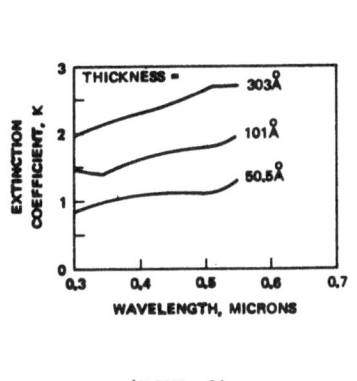

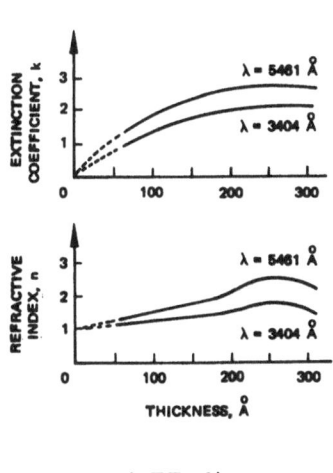

(REF. 1)

(REF. 1)

REFERENCES:

1. D. Malé and J. Trompette, J. Phys. Radium, <u>18</u>, 128-130, (1957).

2. Degussa, "Edelmetall - Taschenbuch," Degussa, Frankfurt/Main, (1967).

PLATINUM

OPTICAL MATERIALS PROPERTIES
DATA SHEET

MATERIAL: __PLATINUM__
__(Film)__

INTRODUCTION: This data sheet contains optical data for platinum film as well as thermal, physical, and mechanical properties for bulk platinum.

PHYSICAL PROPERTIES, (298°K)

Density, (g/cm^3) ____21.45____

Melting/Softening
Temp. (°K) / ____2046____

Solubility in Water,
(g./100 g. H$_2$O ____<0.01____

MECHANICAL PROPERTIES*, (298°K)

Young's Modulus, (psi) __2.1 x 10^7__

Hardness, (Vickers)____37____

THERMAL PROPERTIES, (298°K)

Linear Expansion
Coeff., (°K)$^{-1}$ ____8.9 x 10^{-6}____

Thermal Conductivity
(10^{-4} cal/(cm sec °K) ____1660____

Specific Heat,
(cal/g)/°K ____0.032____

*Annealed bulk metal

OPTICAL PROPERTIES, (298°K)

Dispersion Equation ____Not available____

Transmission Region, (External
Transmittance ≥10% with __2.0__ mm.
thickness) ____Not available____

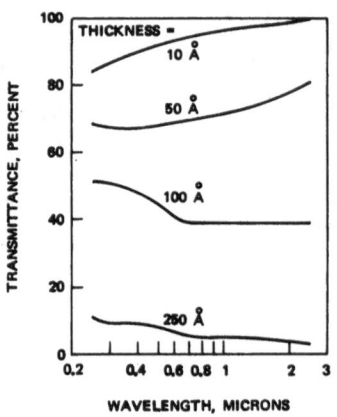

(REF. 1)

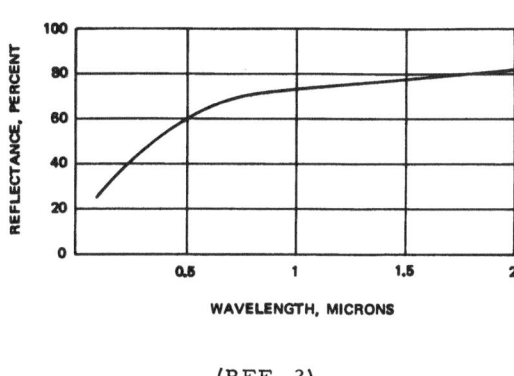

(REF. 2)

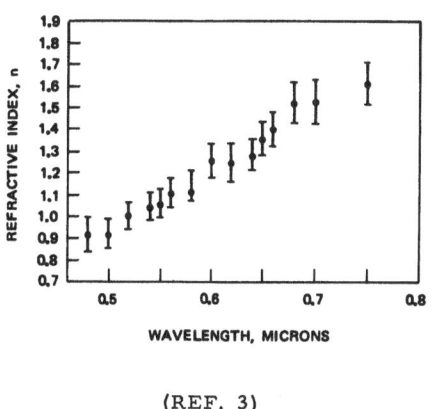

(REF. 3)

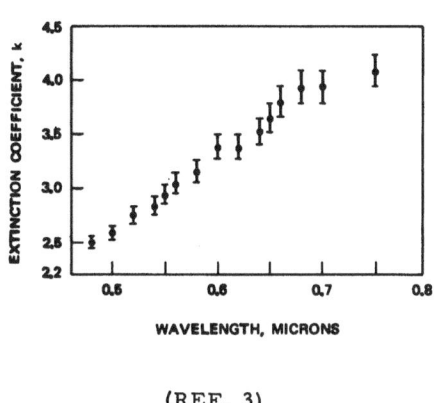

(REF. 3)

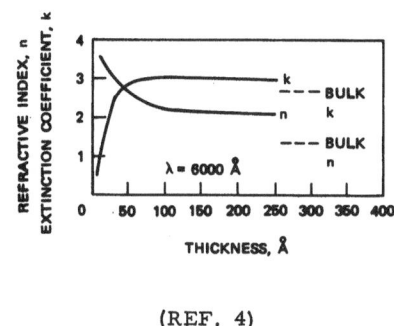

(REF. 4)

REFERENCES:

1. E. Schuch, Ann. der Physik, <u>13</u>, 297-314, (1932).

2. Degussa, "Edelmetall-Taschenbuch," Degussa, Frankfurt/Main, (1967).

3. V. L. Rideout and S. H. Wemple, J. Opt. Soc. Am., <u>56</u>, 749-751, (1966).

4. O. S. Heavens, "Optical Properties of Thin Solid Films," Academic Press, N. Y., (1955).

POTASSIUM BROMIDE

OPTICAL MATERIALS PROPERTIES
DATA SHEET

MATERIAL: POTASSIUM BROMIDE

INTRODUCTION: This data sheet contains information for single crystal potassium bromide.

PHYSICAL PROPERTIES, (298°K)

Density, (g/cm³) ___ 2.75 ___

Melting/Softening
Temp. (°K) ___ 1003 ___

Solubility in Water,
(g./100 g. H_2O) ___ 65.2 (293°K) ___

MECHANICAL PROPERTIES, (298°K)

Young's Modulus, (psi) ___ 3.9×10^{-6} ___

Hardness, (Knoop) ___ 7.0 <100> (200g.) ___

THERMAL PROPERTIES, (298°K)

Linear Expansion
Coeff. (°K⁻¹) ___ 43×10^{-6} ___

Thermal Conductivity
(10^{-4}cal/(cm sec °K) ___ 115 (319°K) ___

Specific Heat,
(cal/g)/°K ___ 0.104 (273°K) ___

OPTICAL PROPERTIES, (298°K)

Dispersion Equation:

$$n^2 = 2.3618102 - 0.00058072\lambda^2$$
$$+ \frac{0.02305269}{\lambda^2 - 0.0245381}$$

(between 0.4 and 0.71μ)

Transmission Region,
(External Transmittance
≥10% with ___ 2.0 ___ mm.
thickness) ___ 0.25 - 35μ ___

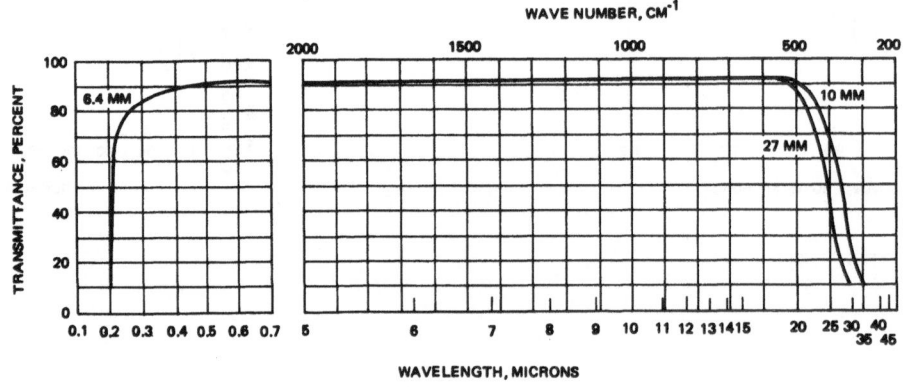

(REF. 1)

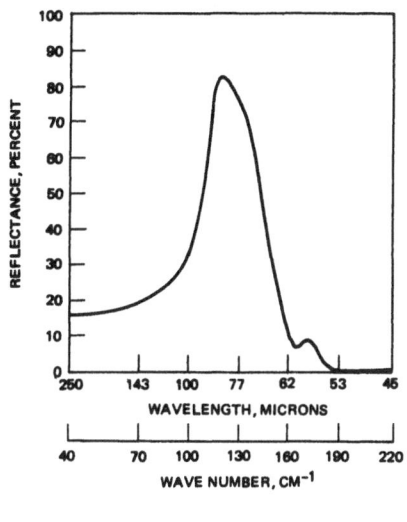

(REF. 2)

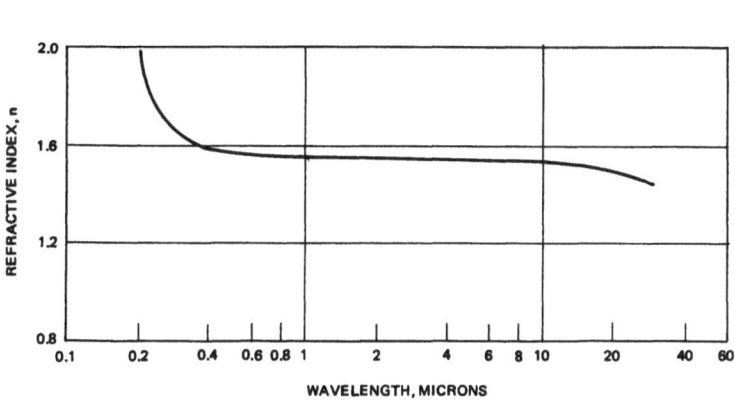

(REF. 3)

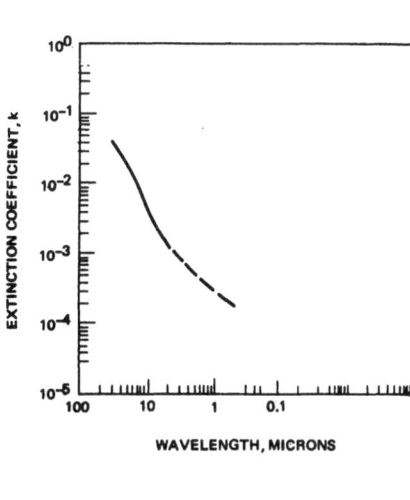

(REF. 4)

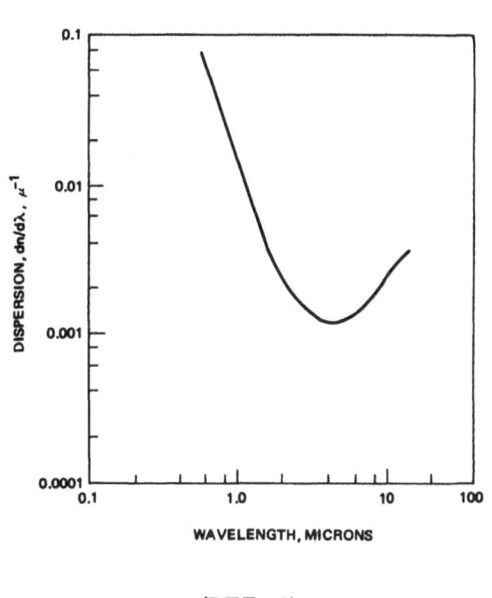

(REF. 5)

REFERENCES:

1. A. Smakula, et al., "Harshaw Optical Crystals", The Harshaw Chemical Co., Cleveland, (1967).

2. C. H. Perry, et al., M.I.T. Cambridge Res. Lab. Electronics, Report No. QPR-91, (1968).

3. A. Smakula, Opt. Acta, <u>9</u>, 205-222, (1962).

4. J. C. Owens, Phys. Rev., <u>181</u>, 1228-1236, (1969).

5. W. L. Wolfe, Ed. "Handbook of Military Infrared Technology", U.S. Government Printing Office, Washington, D.C., (1965).

POTASSIUM CHLORIDE

OPTICAL MATERIALS PROPERTIES
DATA SHEET

MATERIAL: <u>POTASSIUM</u>
<u>CHLORIDE</u>

INTRODUCTION: This data sheet summarizes properties of single crystal potassium chloride.

PHYSICAL PROPERTIES, (298°K)

Density, (g/cm^3)   1.99

Melting/Softening
Temp. (°K)   1049

Solubility in Water,
(g./100 g. H$_2$0)   34.35 (293°K)

MECHANICAL PROPERTIES, (298°K)

Young's Modulus, (psi)   4.30 x 10^{-6}

Hardness, (Knoop)   9.3 <100> (200g)

THERMAL PROPERTIES, (298°K)

Linear Expansion
Coeff. (°K^{-1})   36 x 10^{-6}

Thermal Conductivity
(10^{-4}cal/(cm sec °K)   156   (315°K)

Specific Heat,
(cal/g)/°K   0.162 (273°K)

OPTICAL PROPERTIES, (298°K)

Dispersion Equation:
(for the ultraviolet and visible)

$$n^2 = a^2 + \frac{M_1}{\lambda^2 - \lambda_1^2} + \frac{M_2}{\lambda^2 - \lambda_2^2} - k\lambda^2 - h\lambda^4, \ n^2 = b^2$$

$$+ \frac{M_1}{\lambda^2 - \lambda_1^2} - \frac{M_2}{\lambda^2 - \lambda_2^2} + \frac{M_3}{\lambda_3^2 - \lambda^2}$$

a^2 = 2.174967   k = 0.000513495

M_1 = 0.008344206   h = 0.06167587

λ_1^2 = 0.0119082 b^2 = 3.866619

M_2 = 0.00698382 M_3 = 5569.715

λ_2^2 = 0.0255550 λ_3^2 = 3292.47

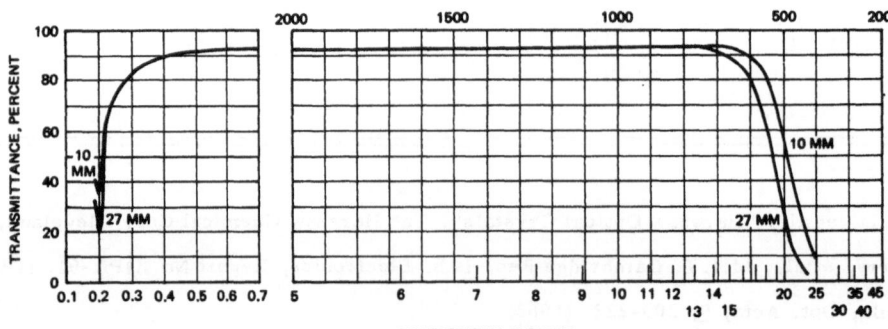

(REF. 1)

58

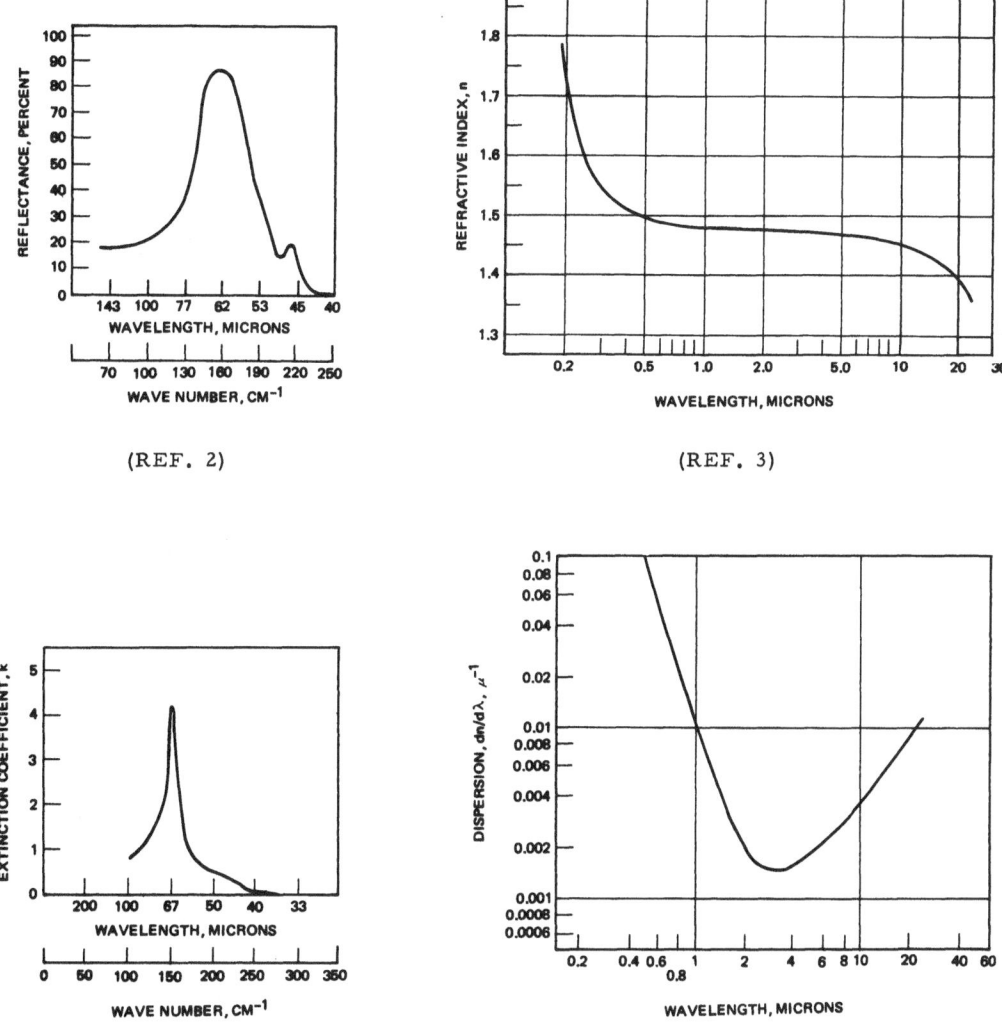

(REF. 2)

(REF. 3)

(REF. 4)

(REF. 5)

REFERENCES:

1. A.Smakula, et al., "Harshaw Optical Crystals", The Harshaw Chemical Co., Cleveland, (1967).

2. C.H.Perry, et al., MIT - Cambridge Res. Lab. Electronics, Report No. QPR-91, (1968).

3. J.A.Mauro, "Optical Engineering Handbook", General Electric Co., Scranton, Pa., (1963).

4. J.N.Plendl and P.J.Gielisse, Appl. Optics, 3, 943-949, (1964).

5. A. Smakula, Opt. Acta, 9, 205-222, (1962).

POTASSIUM IODIDE

OPTICAL MATERIALS PROPERTIES
DATA SHEET

MATERIAL: <u>POTASSIUM
IODIDE</u>

INTRODUCTION: This data sheet presents properties of single crystal potassium iodide.

PHYSICAL PROPERTIES, (298°K)

Density, (g/cm^3)   3.13

Melting/Softening
Temp. (°K)   996

Solubility in Water,
(g./100 g. H$_2$0)   144.5 (293°K)

MECHANICAL PROPERTIES, (298°K)

Young's Modulus, (psi)   4.57 x 10^{-6}

Hardness, (Knoop)   5

THERMAL PROPERTIES, (298°K)

Linear Expansion
Coeff. (°K^{-1})   42.6 x 10^{-6} (313°K)

Thermal Conductivity
(10^{-4}cal/(cm sec °K)   50.1

Specific Heat,
(cal/g)/°K   0.075 (270°K)

OPTICAL PROPERTIES, (298°K)

Dispersion Equation: Not available

Transmission Region,
(External Transmittance
≥10% with ___2.0___ mm.
thickness)   0.25 - 45μ

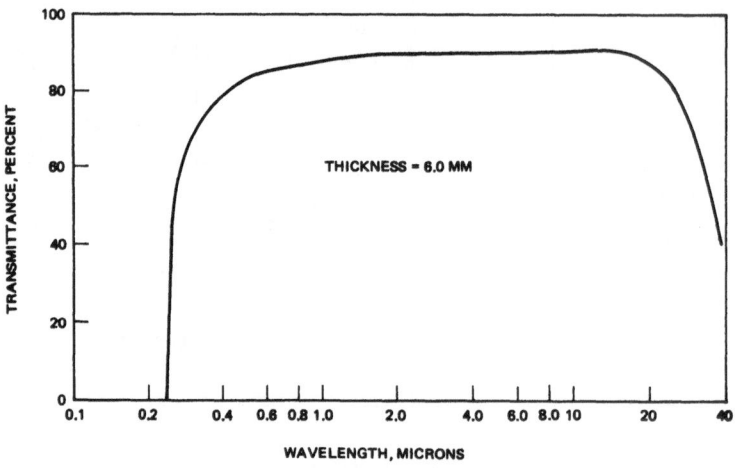

(REF. 1)

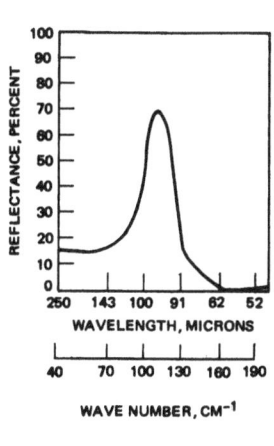

(REF. 2)

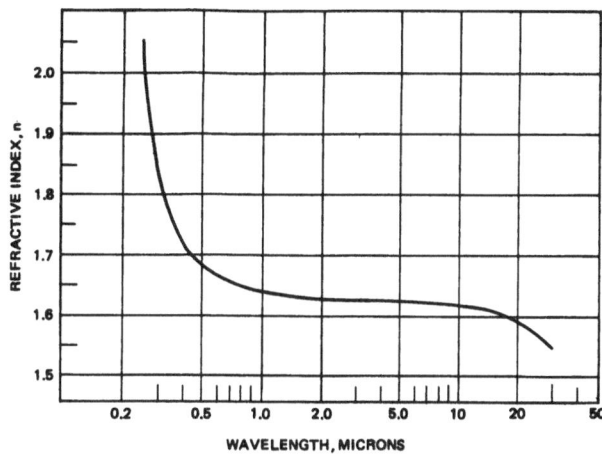

(REF. 3)

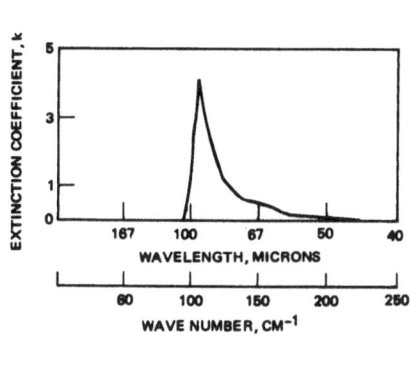

(REF. 4)

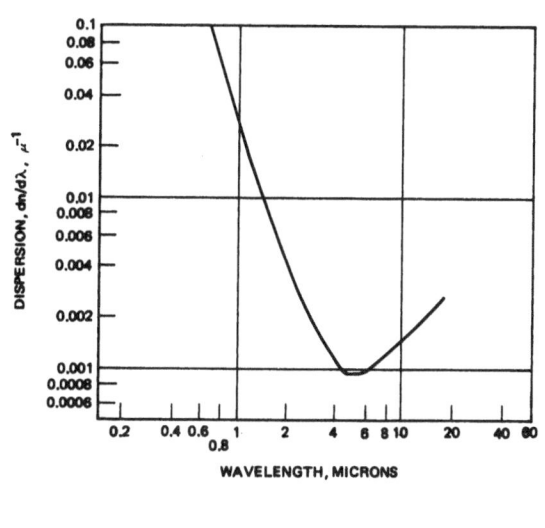

(REF. 5)

REFERENCES:

1. Isomet Corp., "Optical Crystals", Bull. No. 1101, (1965).

2. C. H. Perry, et al., MIT, Cambridge Res. Lab. of Electronics, Report No. QPR-91, (1968).

3. J. A. Mauro, "Optical Engineering Handbook", General Electric Co., Scranton, Pa., (1963).

4. A. Hadni, et al., Appl. Optics, 7, 161-165, (1968).

5. A. Smakula, Opt. Acta, 9, 205-222, (1962).

SELENIUM
(Amorphous)

OPTICAL MATERIALS PROPERTIES
DATA SHEET

MATERIAL: SELENIUM
(Amorphous
Film)

INTRODUCTION: This data sheet presents properties of amorphous, vacuum-deposited selenium film.

PHYSICAL PROPERTIES, (298°K)

Density, (g/cm^3) 4.25

Softening Temp. (°K) 313 - 323

Solubility in Water,
(g. /100 g. H$_2$0) <0.005

MECHANICAL PROPERTIES, (298°K)

Young's Modulus, (psi) not available

Hardness, (Knoop) not available

THERMAL PROPERTIES, (298°K)

Linear Expansion
Coeff. (°K)$^{-1}$ 37 x 10^{-6}

Thermal Conductivity
(10^{-4}cal/(cm sec °K) 31

Specific Heat,
(cal/g)/°K 0.08

OPTICAL PROPERTIES, (298°K)

Dispersion Equation: not available

Transmission Region,
(External Transmittance
≥10% with 2.0 mm.
thickness) 1.0 - 20μ

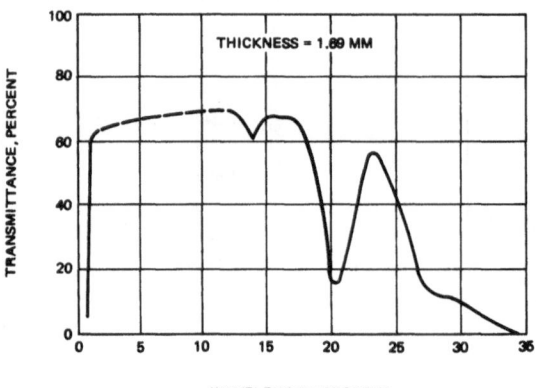

(REF. 1)

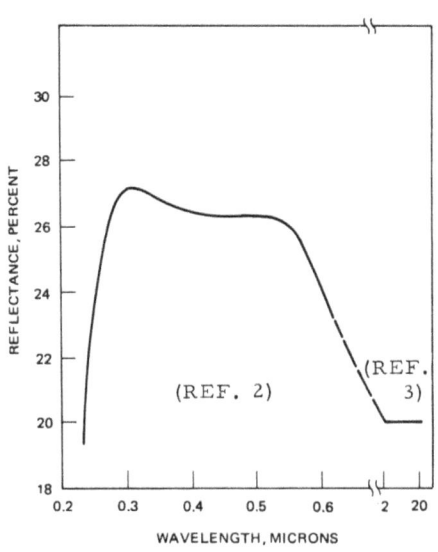

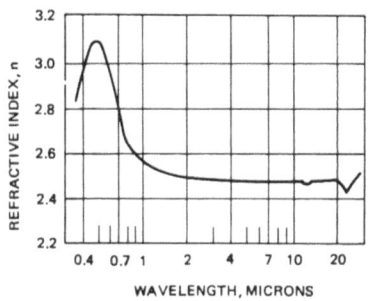

(REF. 4)

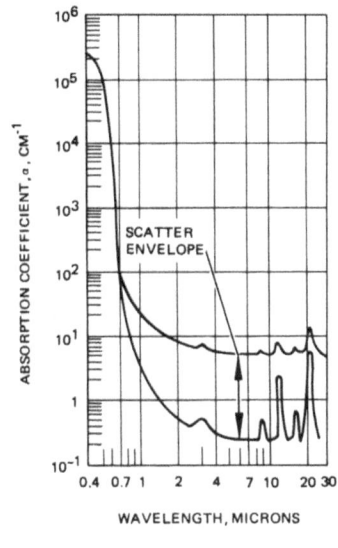

(REF. 4)

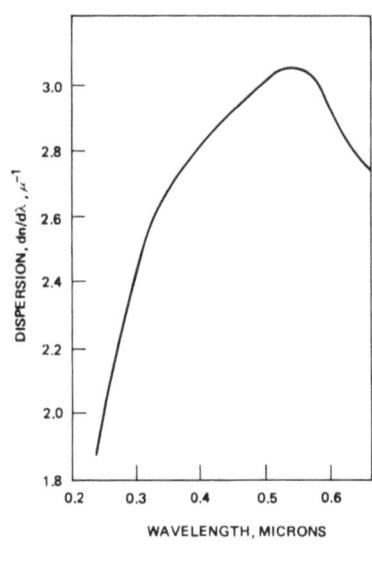

(REF. 2)

REFERENCES:

1. R.S.Caldwell, Special Report on Contract DA 36-039-SC-71131, Purdue University, (1958).

2. W.F.Koehler, et al., J. Opt. Soc. Am., 49, 109-115, (1959).

3. A.Vasko, Czech. J. Phys., 15, 170-177, (1965).

4. H.Gobrecht and A.Tausend, Z.Physik, 161, 205-220, (1961).

SELENIUM
(Hexagonal)

OPTICAL MATERIALS PROPERTIES
DATA SHEET

MATERIAL: SELENIUM -
HEXAGONAL

INTRODUCTION: This data sheet summarizes properties of single crystal hexagonal selenium.

PHYSICAL PROPERTIES, (298°K)

Density, (g/cm^3)　　4.82

Melting/Softening
Temp. (°K)　　490

Solubility in Water,
(g./100 g. H$_2$O)　　<0.005

MECHANICAL PROPERTIES, (298°K)

Young's Modulus, (psi)　　8.4 x 10^6

Hardness, (Mohs)　　2.0

THERMAL PROPERTIES, (298°K)

Linear Expansion
Coeff., (°K)$^{-1}$　　37.9

Thermal Conductivity
(10^{-4} cal/(cm sec °K)　　60

Specific Heat,
(cal/g)/°K)　　0.076

OPTICAL PROPERTIES, (298°K)

Dispersion Equation　　Not available

Transmission Region, (External
Transmittance ≥10% with_____mm.
thickness)　　Not available

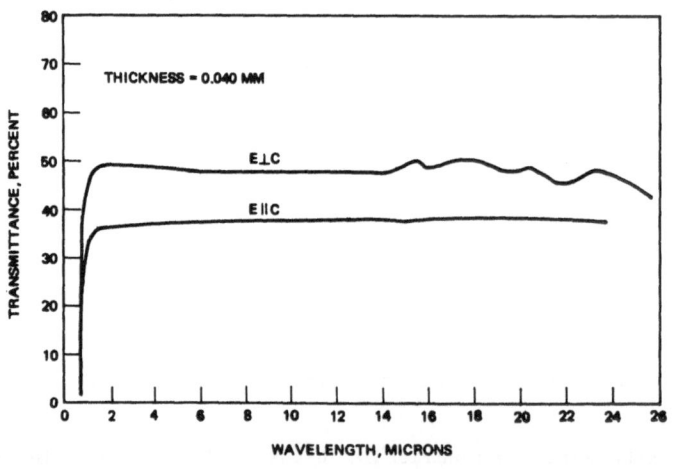

(REF. 1)

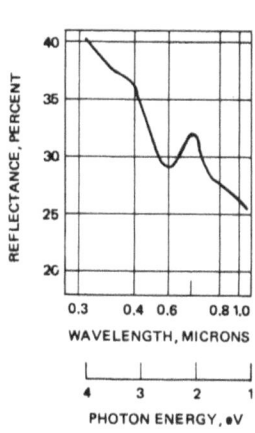

(REF. 2)

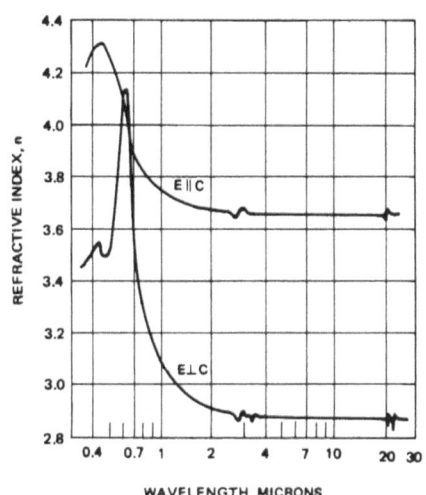

(REF. 3)

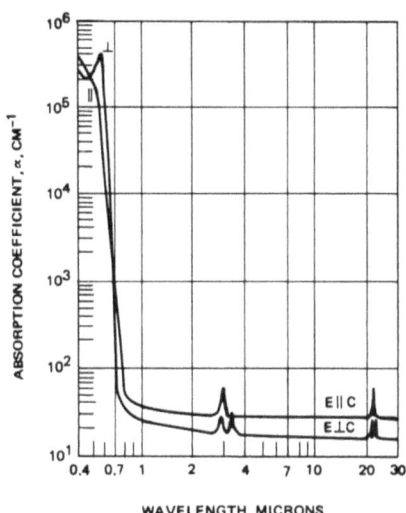

(REF. 3)

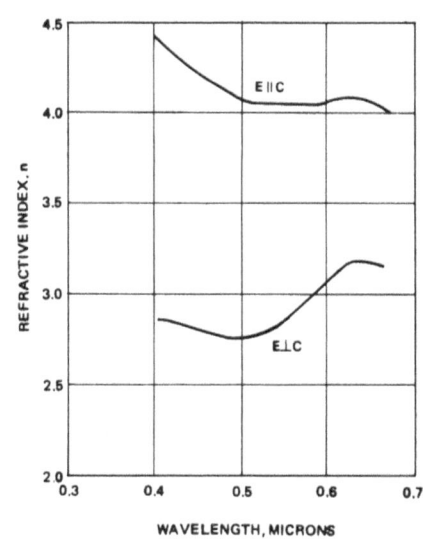

(REF. 4)

REFERENCES:

1. Purdue University, U.S. Govt. Report No. AD 54914, (1954).

2. J. Stuke, Z. Physik, 134, 194-207, (1953).

3. H. Gobrecht and A. Tausend, Z. Physik, 161, 205-220, (1961).

4. C.H. Skinner, Phys. Rev., 9, 148-159, (1917).

65

SILICA
(Crystalline)

OPTICAL MATERIALS PROPERTIES
DATA SHEET

MATERIAL: SILICA-
CRYSTALLINE

INTRODUCTION: This data sheet contains information for high purity crystalline silica.

PHYSICAL PROPERTIES, (298°K)

Density, (g/cm^3) _____ 2.65

Melting/Softening Temp. (°K) ____ 1743

Solubility in Water,
(g./100 g. H$_2$0) _____ < 0.001

MECHANICAL PROPERTIES, (298°K)*

Young's Modulus, (psi) ____ (14.1-11.1) x 10

Hardness, (Knoop) _____ 741 (500 g.)

THERMAL PROPERTIES, (298°K)*

Linear Expansion
Coeff. (°K^{-1}) _____ (7.97-13.37) x 10^{-6}

Thermal Conductivity
(10^{-4}cal/(cm sec °K) _____ 255 - 148

Specific Heat,
(cal/g)/°K _____ 0.188

*Double values for E ∥ C and E ⊥ C, respectively.

OPTICAL PROPERTIES, (298°K)

Dispersion Equation: _____ not available

Transmission Region,
(External Transmittance
≥10% with __ 2.0 __ mm.
thickness) _____ 0.12 - 4.5μ

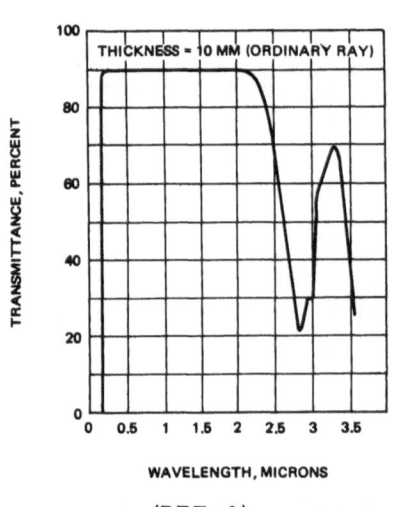

(REF. 1)

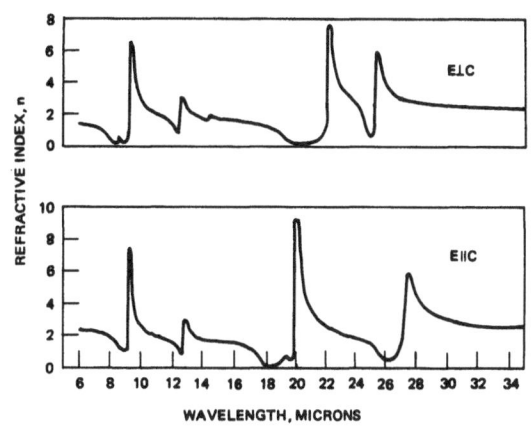

(REF. 2)

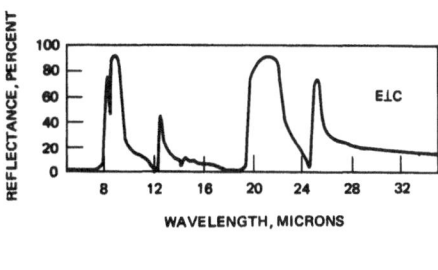

(REF. 2)

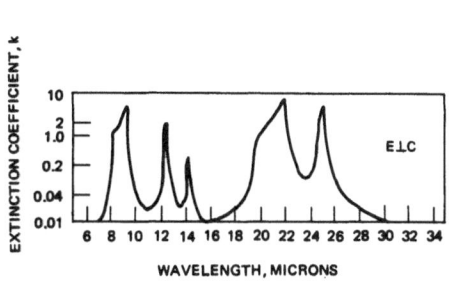

(REF. 2)

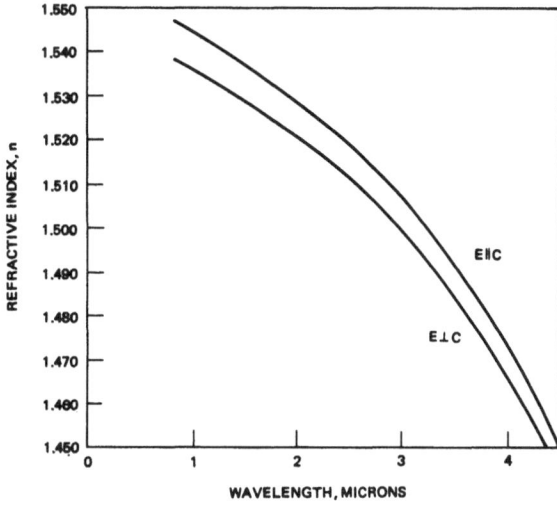

(REF. 3)

REFERENCES:

1. P. Billard, Acta Electronica, <u>6</u>, 75 - 169, (1962).

2. W. G. Spitzer and D. A. Kleinman, Phys. Rev., <u>121</u>, 1324 - 1335, (1961).

3. American Institute of Physics Handbook, McGraw-Hill Book Co., Inc., New York, (1957).

SILICA
(Fused)

OPTICAL MATERIALS PROPERTIES
DATA SHEET

MATERIAL: <u>SILICA-
FUSED</u>

INTRODUCTION: This data sheet summarizes properties of high purity fused silica.

PHYSICAL PROPERTIES, (298°K)

Density, (g/cm^3) 2.20

Melting/Softening
Temp. (°K)   1943

Solubility in Water,
$(g./100 g. H_2O)$ <0.001

MECHANICAL PROPERTIES, (298°K)

Young's Modulus, (psi) 10.6×10^6

Hardness, (Knoop)   461 (200 g.)

THERMAL PROPERTIES, (298°K)

Linear Expansion
Coeff. $(°K^{-1})$ 0.55×10^{-6}

Thermal Conductivity
$(10^{-4} cal/(cm\ sec\ °K)$ 33.0

Specific Heat,
$(cal/g)/°K$ 0.18

OPTICAL PROPERTIES, (298°K)

Dispersion Equation:

$$n^2 = 2.978645 + \frac{0.008777808}{\lambda^2 - 0.010609}$$

$$- \frac{84.06224}{96.00000 - \lambda^2}$$

Transmission Region,
(External Transmittance
≥10% with   2.0   mm.
thickness)   0.12 - 4.5μ

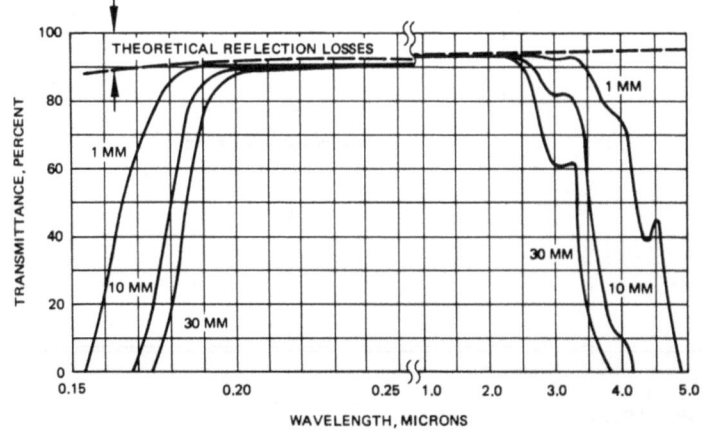

(REF. 1)

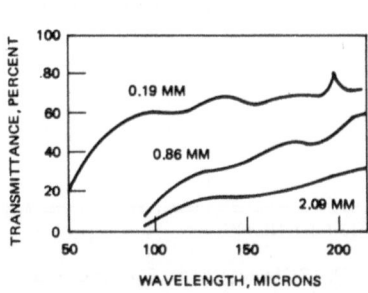

(REF. 2)

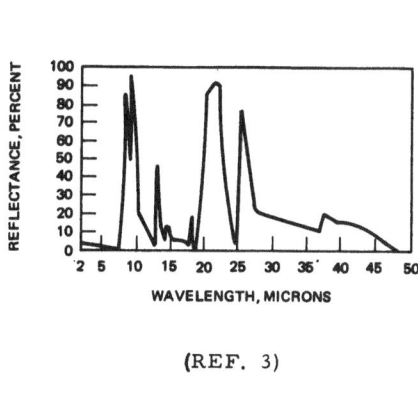

(REF. 3)

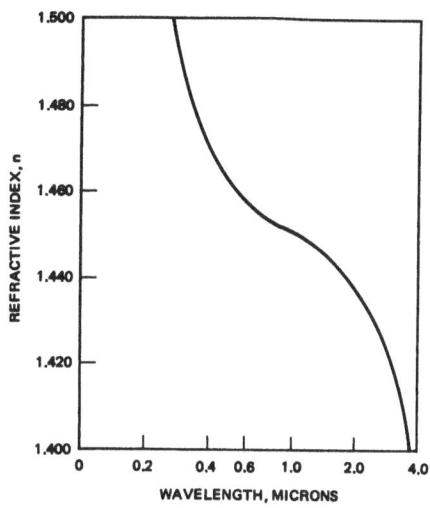

(REF. 4)

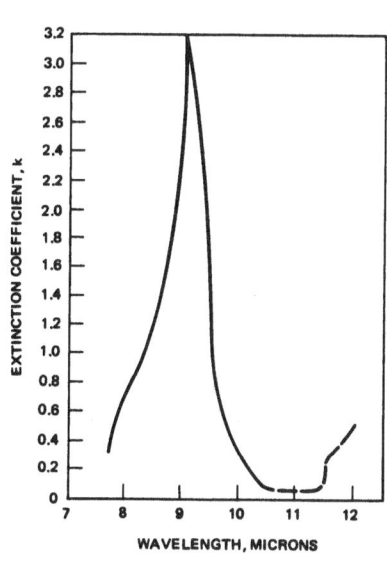

(REF. 5)

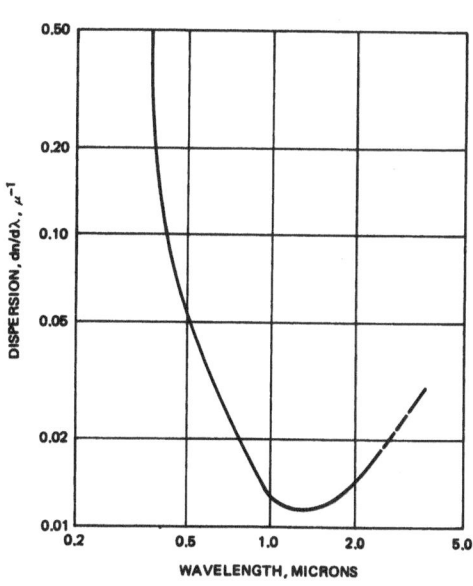

(REF. 6)

REFERENCES:

1. Amersil, Inc. Data Sheet.

2. C. H. Cartwright, Z. Physik, 90, 480-488, (1934).

3. D. E. McCarthy, Appl. Optics, 2, 591-595, (1963).

4. I. H. Malitson, J. Opt. Soc. Am., 55, 1205-1209, (1965).

5. G. W. Cleek, Appl. Optics, 5, 771-776, (1966).

6. W. S. Rodney and R. J. Spindler, J. Res. NBS, 53, 185-189, (1964).

SILICON

OPTICAL MATERIALS PROPERTIES
DATA SHEET

MATERIAL: <u>SILICON</u>

INTRODUCTION: <u>This data sheet covers properties of single crystal silicon.</u>

PHYSICAL PROPERTIES, (298°K)

Density, (g/cm^3) _____ 2.33 _____

Melting/Softening
Temp. (°K) _____ 1693

Solubility in Water,
(g. /100 g. H$_2$0) _____ < 0.005 _____

MECHANICAL PROPERTIES, (298°K)

Young's Modulus, (psi) _____ 19.0 x 10^6 _____

Hardness, (Knoop) _____ 1100 - 1400 _____

THERMAL PROPERTIES, (298°K)

Linear Expansion
Coeff. (°K^{-1}) _____ 4.7 x 10^{-6}

Thermal Conductivity
(10^{-4}cal/(cm sec °K) _____ 3900 (313°K) _____

Specific Heat,
(cal/g)/°K _____ 0.168 _____

OPTICAL PROPERTIES, (298°K)

Dispersion Equation:

$$n = 3.41696 + 0.138497 \, L + 0.013924 \, L^2$$
$$- 0.0000209 \, \lambda^2 + 0.000000148 \, \lambda^4$$

where: $L = (\lambda^2 - 0.028)^{-1}$

Transmission Region,
(External Transmittance
≥10% with ___2.0___ mm.
thickness) _____ 1.2 - 15µ

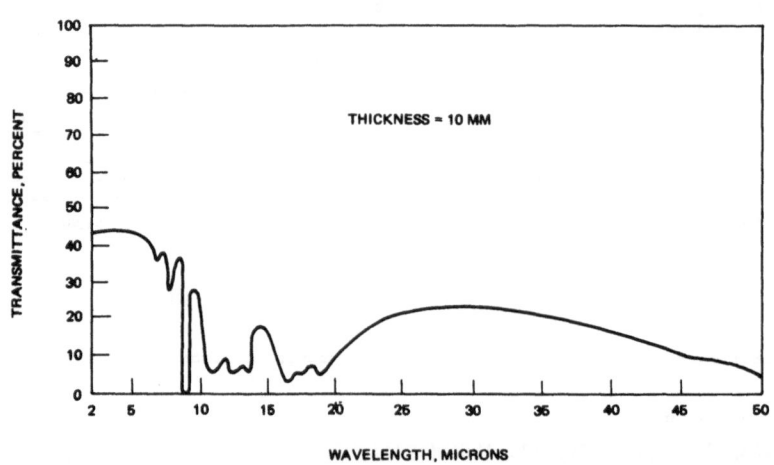

(REF. 1)

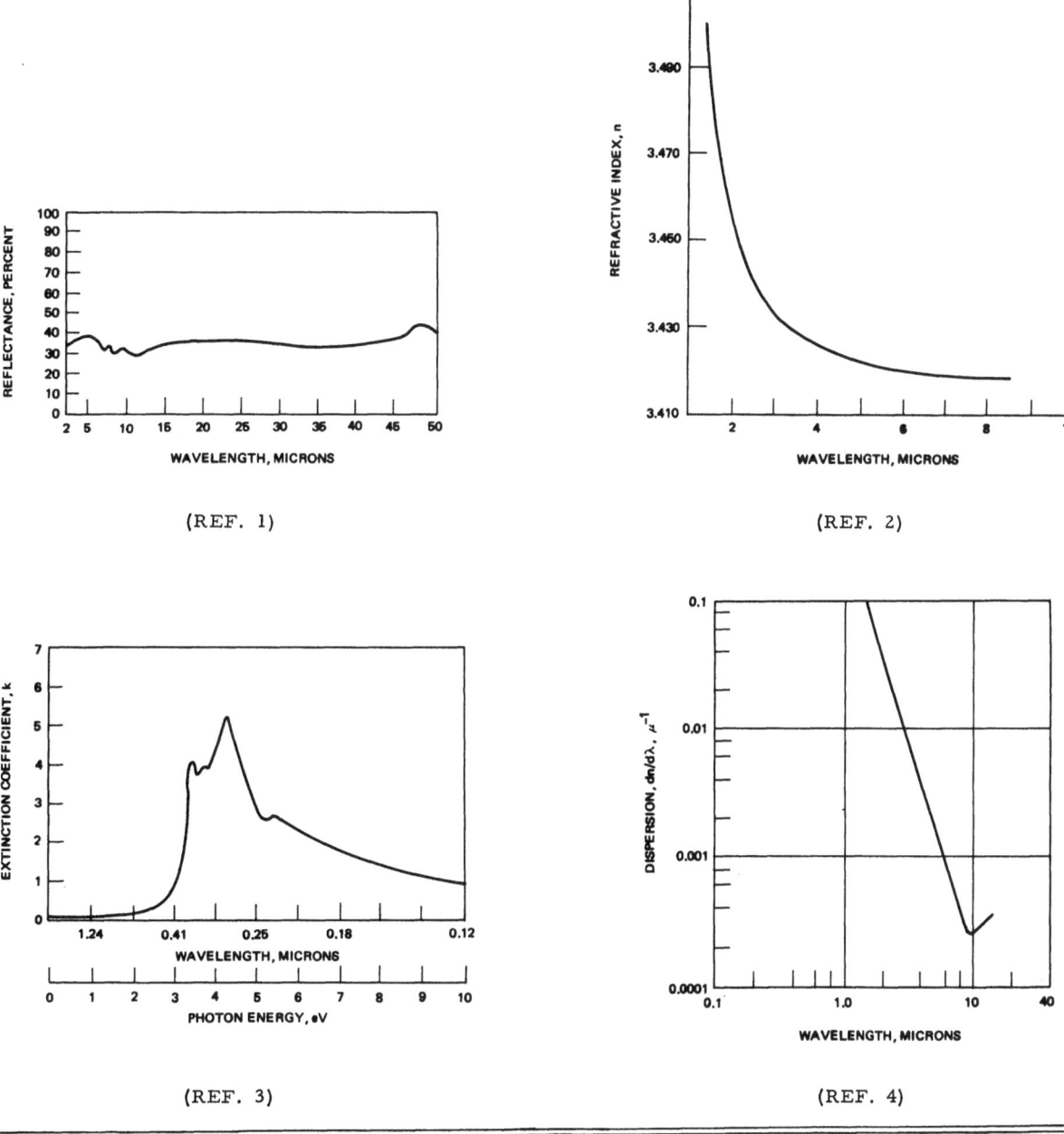

(REF. 1)

(REF. 2)

(REF. 3)

(REF. 4)

REFERENCES:

1. D. E. McCarthy, Appl. Optics, 2, 591-603, (1963).

2. C. D. Salzberg and J. J. Villa, J. Opt. Soc. Am., 47, 244-246, (1957).

3. H. W. Verleur, J. Opt. Soc. Am., 58, 1356-1364, (1968).

4. S. S. Ballard, et al., "Optical Materials for Infrared Instrumentation", Report No. AD217367, (1959).

SILICON CARBIDE

OPTICAL MATERIALS PROPERTIES
DATA SHEET

MATERIAL: SILICON
CARBIDE
(Type 6H)

INTRODUCTION: This data sheet contains information for single hexagonal silicon carbide crystals, type 6H.

PHYSICAL PROPERTIES, (298°K)

Density, (g/cm^3)   3.21

Melting/Softening
Temp. (°K)   3103

Solubility in Water,
(g./100 g. H_2O)   <0.01

MECHANICAL PROPERTIES, (298°K)

Young's Modulus, (psi) 56×10^6

Hardness, (Knoop)   2130 - 2755

THERMAL PROPERTIES, (298°K)

Linear Expansion
Coeff., (°K)$^{-1}$ $\sim 4 \times 10^{-6}$

Thermal Conductivity
(10^{-4} cal/cm sec °K)   1000

Specific Heat,
(cal/g)/°K   0.165

OPTICAL PROPERTIES, (298°K)

Dispersion Equation   Not available

Transmission Region, (External
Transmittance ≥10% with_____mm.
thickness)   Not available

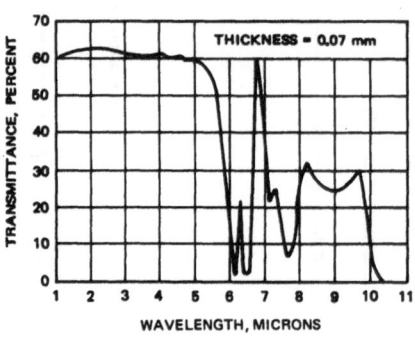

(REF. 1)

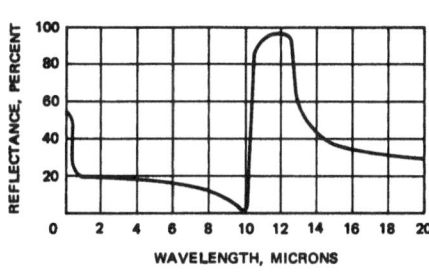

(REF. 2, 3)

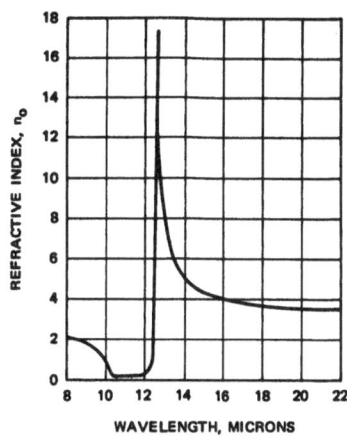

(REF. 2)

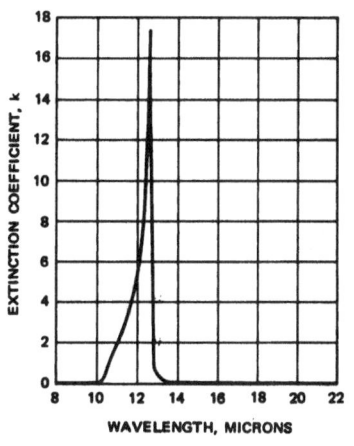

(REF. 2)

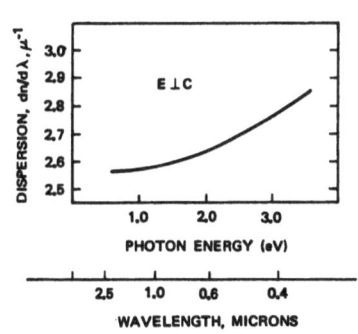

(REF. 4)

REFERENCES:

1. H.G. Lipson, Conf. on Silicon Carbide, Boston, 1959, Silicon Carbide - A High Temperature Semiconductor; Proc., Ed. J. R. O'Connor and J. Smittens, Pergamon Press, N.Y., (1960).

2. W. G. Spitzer, et al, Phys. Rev., 113, 127-132, (1959).

3. H.R. Philipp and E.A. Taft, Conf. on Silicon Carbide, Boston, 1959, Silicon Carbide - A High Temperature Semiconductor, Proc. Ed., J. R. O'Connor and J. Smittens, Pergamon Press, N.Y., (1960).

4. W.J. Choyke and L. Patrick, J. Opt. Soc. Am., 58, 377-379, (1968).

SILVER

OPTICAL MATERIALS PROPERTIES   MATERIAL: <u>SILVER (Film)</u>
DATA SHEET

INTRODUCTION: This data sheet presents properties of vacuum-evaporated silver film, generally on a glass substrate.

PHYSICAL PROPERTIES, (298°K)

Density, (g/cm^3)   10.5

Melting/Softening
Temp. (^{O}K)   1234

Solubility in Water,
(g./100 g. H$_2$O)   < 0.001

MECHANICAL PROPERTIES, (298°K)

Young's Modulus, (psi)   8.27 x 10^5

Hardness, (Rockwell F)   91 (bulk)

THERMAL PROPERTIES, (298°K)

Linear Expansion
Coeff., (^{O}K)$^{-1}$ 19.1 x 10^{-6}

Thermal Conductivity
10^{-4} cal/(cm sec ^{O}K)   1.0 x 10^4

Specific Heat,
(cal/g)/^{O}K   0.056

OPTICAL PROPERTIES, (298°K)

Dispersion Equation   Not available

Transmission Region, (External
Transmittance ≥10% with _____ mm.
thickness)   Not available

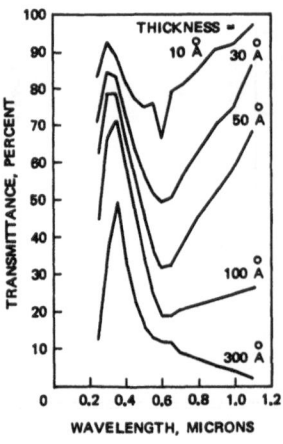

(REF. 1)

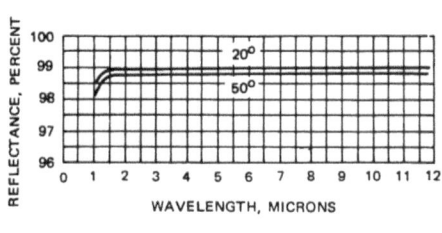

(REF. 2)

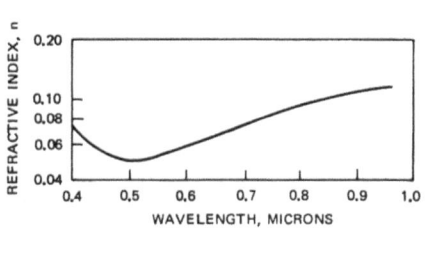

(REF. 3)

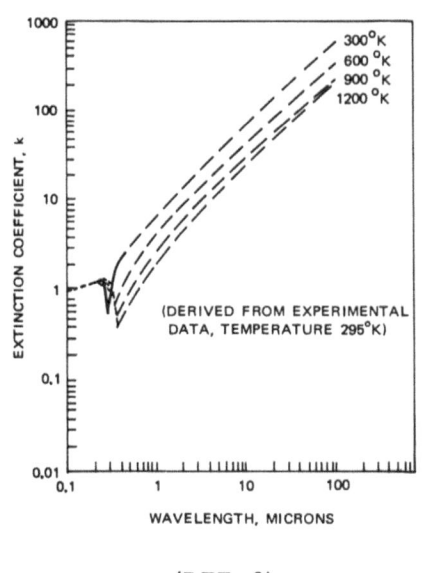

(REF. 3)

(REF. 4)

REFERENCES:

1. F. Goos, Z. Phys., <u>100</u>, 95-112, (1936).

2. D.M. Gates, et al, J. Opt. Soc. Am., <u>48</u>, 88-89, (1958).

3. A.F. Grenis, U.S. Government Report No. AMRA TR 67-02, (1967).

4. L.G. Schulz and F. R. Tangherlini, J. Opt. Soc. Am., <u>44</u>, 362-368, (1965).

SILVER BROMIDE

OPTICAL MATERIALS PROPERTIES
DATA SHEET

MATERIAL: <u>SILVER</u>
<u>BROMIDE</u>

INTRODUCTION: <u>This data sheet contains information for single crystal silver bromide.</u>

PHYSICAL PROPERTIES, (298°K)

Density, (g/cm^3)   6.47

Melting/Softening
Temp. (°K)   705

Solubility in Water,
(g./100 g. H$_2$O)   12.6 x 10^{-6} (293°K)

MECHANICAL PROPERTIES, (298°K)

Young's Modulus, (psi)   4.64 x 10^6

Hardness, (Knoop)   Not available

THERMAL PROPERTIES, (298°K)

Linear Expansion
Coeff., (°K)$^{-1}$ 34.8 x 10^{-6}

Thermal Conductivity
(10^{-4} cal/(cm sec °K   29 (273°K)

Specific Heat,
(cal/g)/°K   0.070

OPTICAL PROPERTIES, (298°K)

Dispersion Equation between 0.54 and 0.65μ:

$$\frac{n^2 - 1}{n^2 + 2} = 0.48484 + \frac{0.10279\lambda^2}{\lambda^2 - 0.090000} - 0.004796\lambda^2$$

Transmission Region, (External
Transmittance ≥10% with <u>2.0</u> mm.
thickness) <u>   0.45 - 35μ</u>

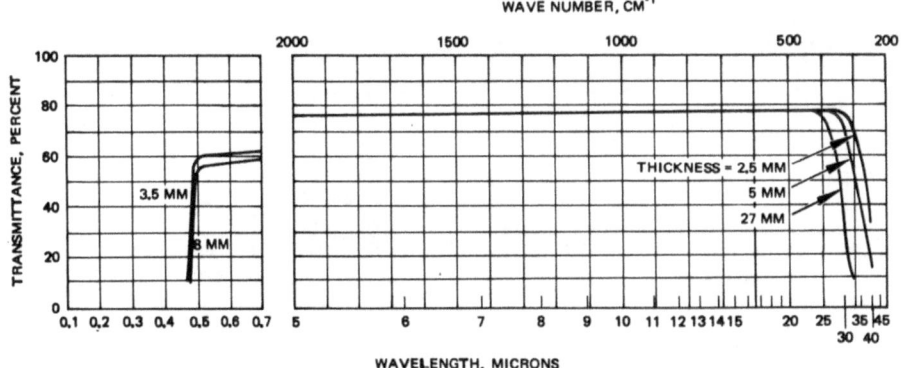

(REF. 1)

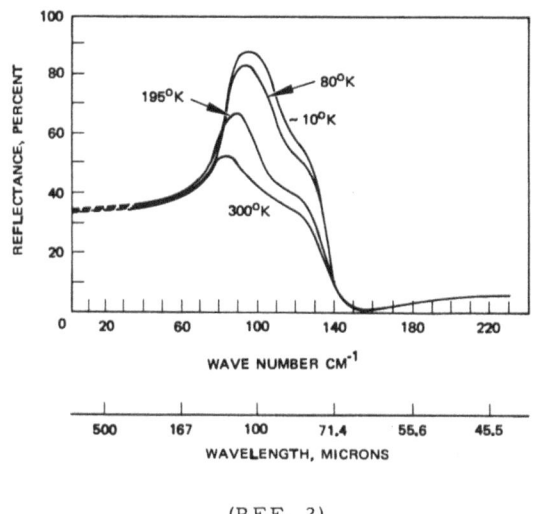

(REF. 2)

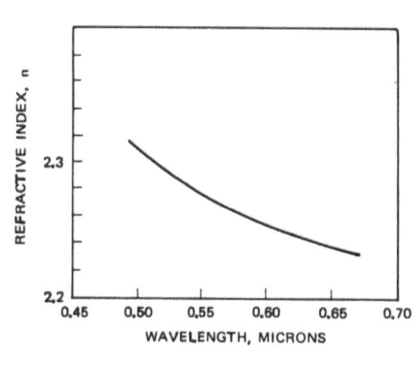

(REF. 3)

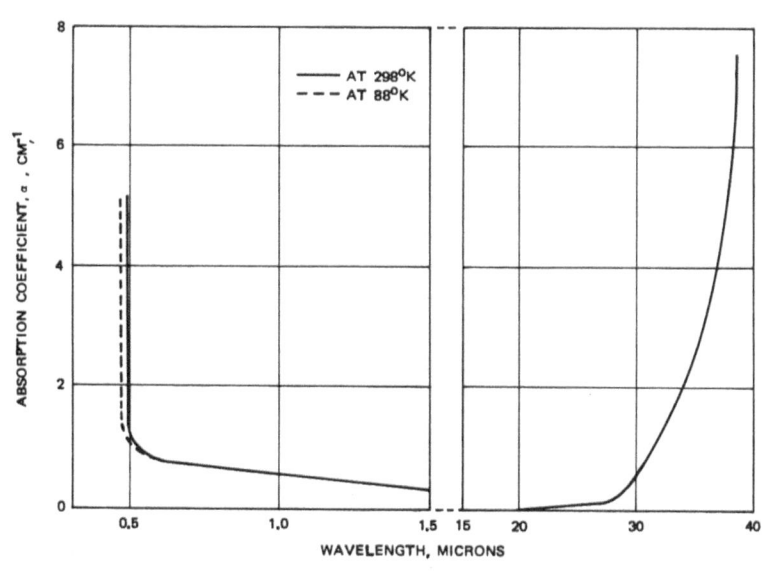

(REF. 4)

REFERENCES:

1. A. Smakula, et al, "Harshaw Optical Crystals," Harshaw Chemical Co., Cleveland, (1967).

2. C.H. Perry, Report No. AD 679 213, (1968).

3. H. Schroeder, Z. Physik, 67, 24-36, (1931).

4. A. Smakula, Report No. AD 663 734, (1967).

SILVER CHLORIDE

OPTICAL MATERIALS PROPERTIES
DATA SHEET

MATERIAL: SILVER CHLORIDE

INTRODUCTION: This data sheet summarizes properties of single crystal silver chloride.

PHYSICAL PROPERTIES, (298°K)

Density, (g/cm^3) _____ 5.56

Melting/Softening
Temp. (°K) _____ 731

Solubility in Water,
(g./100 g. H$_2$O) _____ 1.5 x 10^{-4}

THERMAL PROPERTIES, (298°K)

Linear Expansion
Coeff., (°K)$^{-1}$ _____ 30 x 10^{-6} (293°K)

Thermal Conductivity
(10^{-4} cal/(cm sec °K) _____ 26 (273°K)

Specific Heat,
(cal/g)/°K _____ 0.0848 (273°K)

MECHANICAL PROPERTIES, (298°K)

Young's Modulus, (psi) _____ 2.9 x 10^6

Hardness, (Knoop) _____ 9.5 (200 g)

OPTICAL PROPERTIES, (298°K)

Dispersion Equation

$$n^2 = 4.00804 - 0.00085111\lambda^2$$
$$- 0.00000019762\lambda^4$$
$$+ 0.079086/(\lambda^2 - 0.04584)$$

Transmission Region, (External
Transmittance ≥10% with 2.0 mm.
thickness) _____ 0.4 - 2.8μ

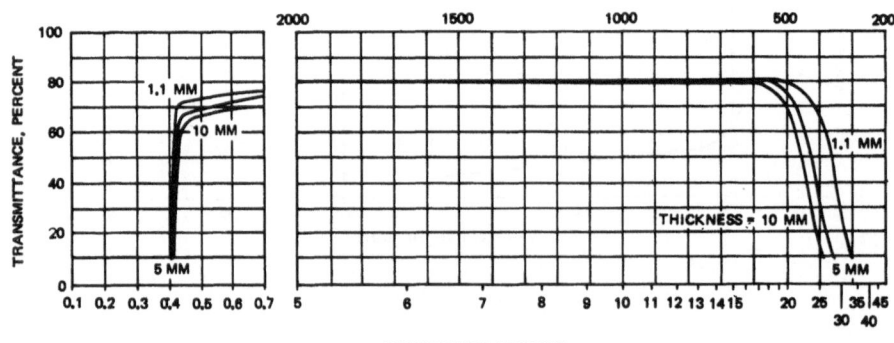

(REF. 1)

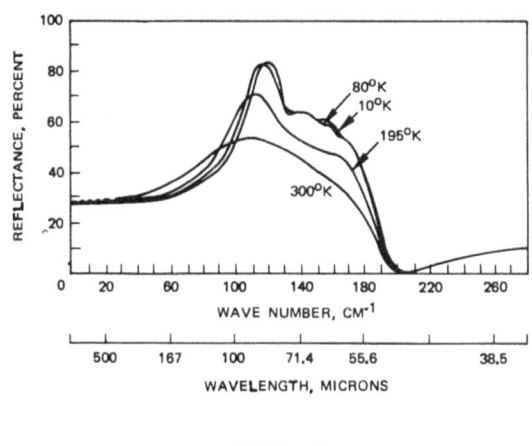

(REF. 2)

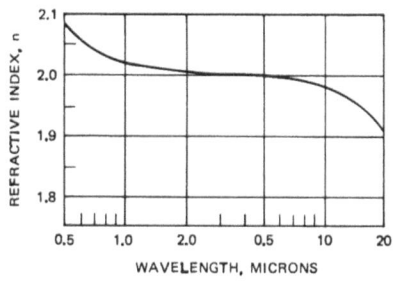

(REF. 3)

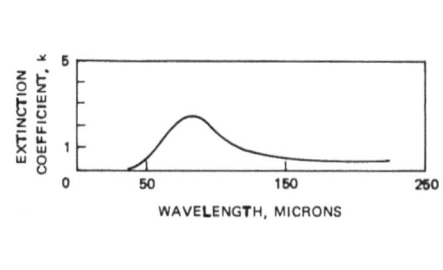

(REF. 4)

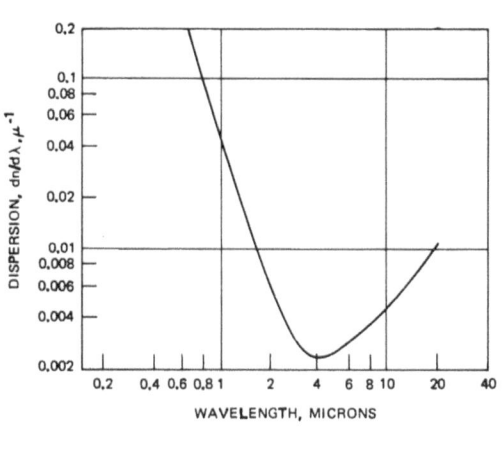

(REF. 5)

REFERENCES:

1. A. Smakula, et al, "Harshaw Optical Crystals," Harshaw Chemical Co., Cleveland, (1967).

2. C.H. Perry, Report No. AD 679 213, (1968).

3. J.A. Mauro, "Optical Engineering Handbook," General Electric Co.. Scranton, Pa., (1963).

4. A. Hadni, et al., J. de Phys., 28, C1-118 - C1-128, (1967).

5. A. Smakula, Opt. Acta, 9, 205-222, (1962).

SODIUM CHLORIDE

<table>
<tr><td>OPTICAL MATERIALS PROPERTIES
DATA SHEET</td><td>MATERIAL: <u>SODIUM</u>
<u>CHLORIDE</u></td></tr>
</table>

INTRODUCTION: <u>This data sheet summarizes properties of single crystal sodium chloride.</u>

PHYSICAL PROPERTIES, (298°K)

Density, (g/cm^3) ___ 3.16

Melting/Softening
Temp. (°K) ___ 1090

Solubility in Water,
(g./100 g. H$_2$O) ___ 36.2

MECHANICAL PROPERTIES, (298°K)

Young's Modulus, (psi) ___ 5.8 x 10^6

Hardness, (Knoop) ___ 17

THERMAL PROPERTIES, (298°K)

Linear Expansion
Coeff., (°K)$^{-1}$ ___ 44 x 10^{-6}

Thermal Conductivity
(10^{-4} cal/(cm sec °K) ___ 155 (289°K)

Specific Heat,
(cal/g)°K ___ 0.204 (273°K)

OPTICAL PROPERTIES, (298°K)

Dispersion Equation ___ Not available

Transmission Region, (External
Transmittance ≥10% with __2.0__ mm.
thickness) ___ 0.21 - 26μ

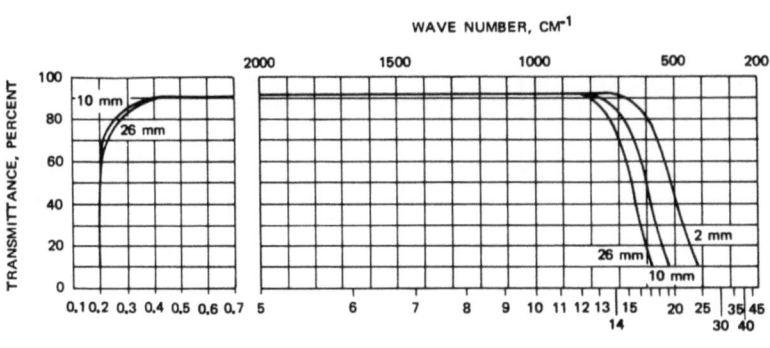

(REF. 1)

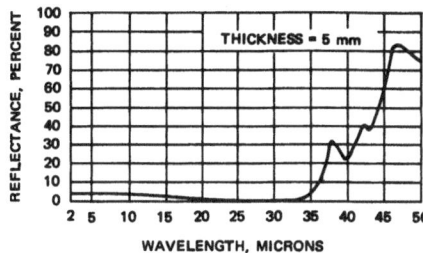

(REF. 2)

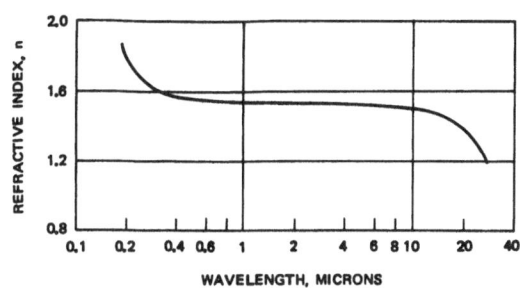

(REF. 3)

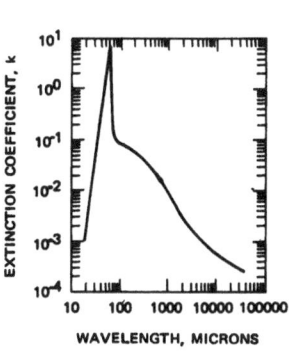

(REF. 4)

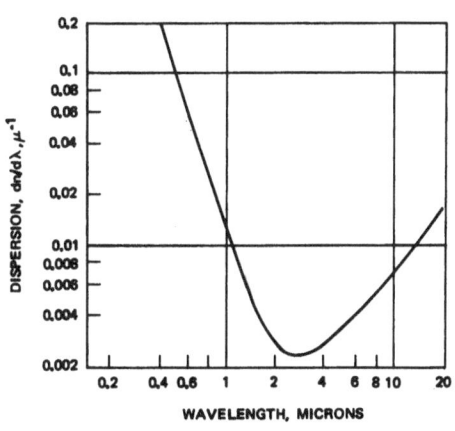

(REF. 3)

REFERENCES:

1. A. Smakula, et al, "Harshaw Optical Crystals," Harshaw Chemical Co., Cleveland, (1967).

2. D.E. McCarthy, Appl. Optics, 2, 591-595, (1963).

3. A. Smakula, Opt. Acta 9, 205-222, (1962).

4. J.C. Owens, Phys. Rev., 181, 1228-1236, (1969).

SODIUM FLUORIDE

OPTICAL MATERIALS PROPERTIES
DATA SHEET

MATERIAL: SODIUM
FLUORIDE

INTRODUCTION: This data sheet summarizes properties of sodium fluoride single crystals.

PHYSICAL PROPERTIES, (298°K)

Density, (g/cm^3)   2.79

Melting/Softening
Temp. (°K)   1270

Solubility in Water,
(g./100 g. H$_2$O)   4.2 (293°K)

MECHANICAL PROPERTIES, (298°K)

Young's Modulus, (psi)   1.41 x 10^7

Hardness, (Knoop)   60

THERMAL PROPERTIES, (298°K)

Linear Expansion
Coeff., (°K)$^{-1}$ 36 x 10^{-6}

Thermal Conductivity
(10^{-4} cal/(cm sec °K)   505

Specific Heat,
(cal/g)/°K   0.26 (273°K)

OPTICAL PROPERTIES, (298°K)

Dispersion Equation   Not available

Transmission Region, (External
Transmittance ≥10% with 2.0 mm.
thickness)   0.19 - 15μ

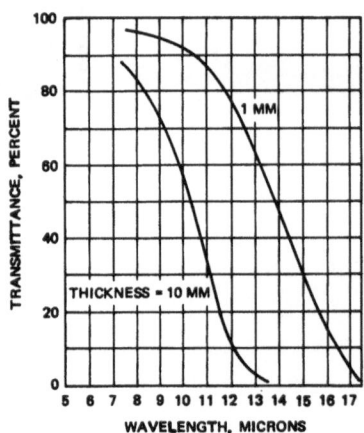

(REF. 1)

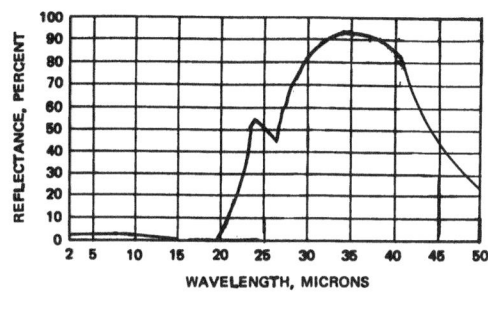

(REF. 2)

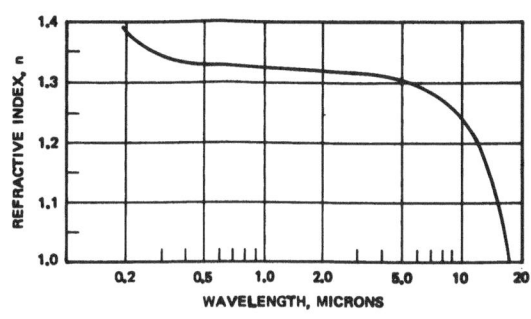

(REF. 3)

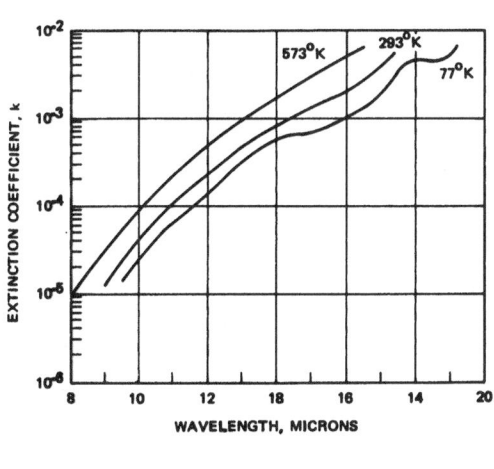

(REF. 4)

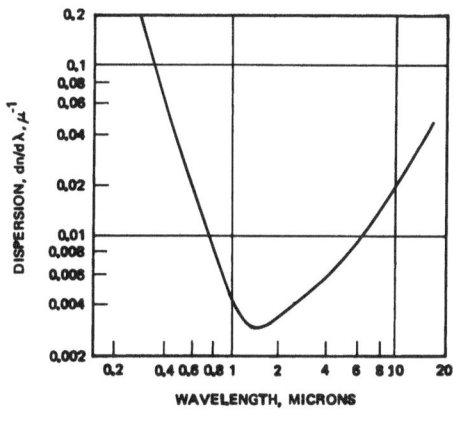

(REF. 5)

REFERENCES:

1. P. Billard, Acta Electronica, 6, 75-169, (1962).

2. D.E. McCarthy, Appl. Optics, 4, 317-320, (1965).

3. J.A. Mauro, "Optical Engineering Handbook," General Electric Co., Scranton, Pa., (1963).

4. M. Klier, Z. Physik, 150, 49-63, (1958).

5. A. Smakula, Opt. Acta, 9, 205-222, (1962).

STRONTIUM TITANATE

OPTICAL MATERIALS PROPERTIES
DATA SHEET

MATERIAL: STRONTIUM
TITANATE

INTRODUCTION: This data sheet summarizes properties of single crystal strontium titanate.

PHYSICAL PROPERTIES, (298°K)

Density, (g/cm³) _____ 5.13

Melting/Softening
Temp. (°K) _____ 2353

Solubility in Water,
(g./100 g. H₂O) _____ <0.01

MECHANICAL PROPERTIES, (298°K)

Young's Modulus, (psi) _____ Not available

Hardness, (Knoop) _____ 595

THERMAL PROPERTIES, (298°K)

Linear Expansion
Coeff., (°K)⁻¹ _____ 9.4×10^{-6}

Thermal Conductivity
(10^{-4} cal/(cm sec °K) _____ 145

Specific Heat,
(cal/g)/°K _____ 0.13

OPTICAL PROPERTIES, (298°K)

Dispersion Equation

$$n = A + BL + CL^2 + D\lambda^2 + E\lambda^4$$

between 1.0 and 5.3µ

A	E	C
2.28355	0.035906	+0.001666

D	E
-0.0061335	-0.00001502

Transmission Region, (External
Transmittance ≥10% with _2.0_ mm.
thickness) _____ 0.39 - 6.8µ

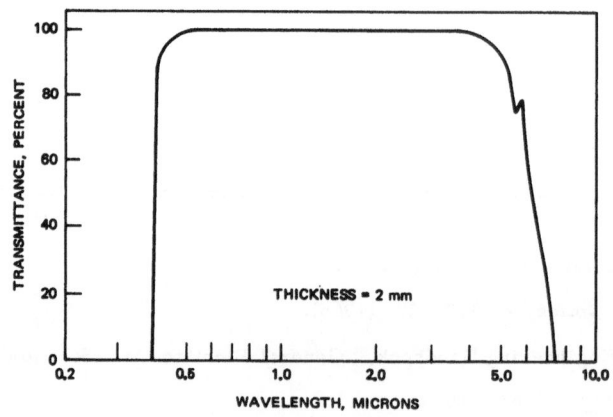

THICKNESS = 2 mm

(REF. 1)

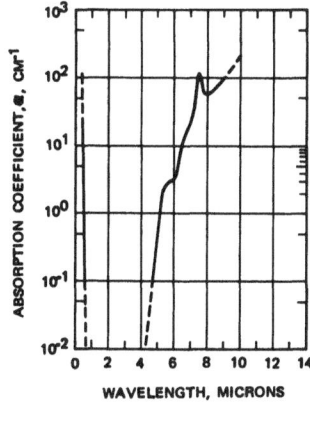

(REF. 2)

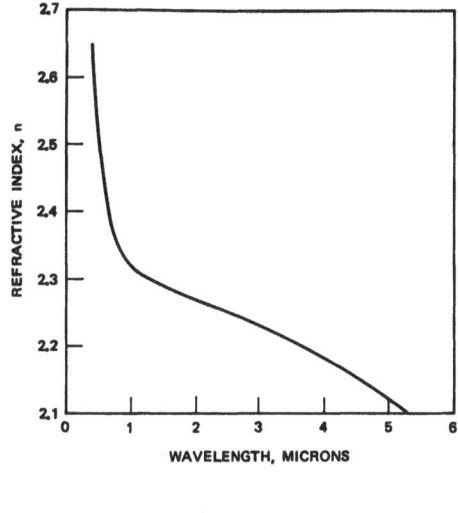

(REF. 3)

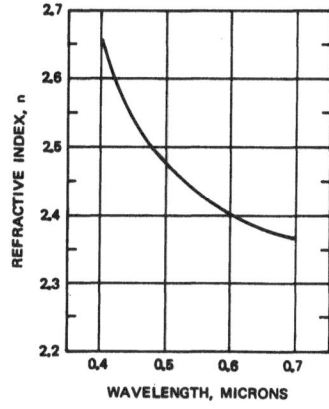

(REF. 4)

(REF. 4)

REFERENCES:

1. M.D. Beals and L. Merker, Materials in Design Eng., <u>51</u>, 12-13, (1960).

2. D.E. McCarthy, Appl. Optics, <u>7</u>, 1997-2000, (1968).

3. S.S. Ballard, et al, Report No. AD 255699, (1961).

4. S.B. Levin, et al, J. Opt. Soc. Am., <u>45</u>, 737-739, (1955).

TELLURIUM
(Film)

<table>
<tr><td colspan="2">OPTICAL MATERIALS PROPERTIES
DATA SHEET</td><td>MATERIAL: <u>TELLURIUM</u>
(Polycrystalline
Film)</td></tr>
</table>

INTRODUCTION: This data sheet summarizes properties of polycrystalline tellurium film.

PHYSICAL PROPERTIES, (298°K)

Density, (g/cm^3)   6.1

Melting/Softening
Temp. (°K)   723

Solubility in Water,
(g./100 g. H$_2$O)   <0.005

MECHANICAL PROPERTIES, (298°K)

Young's Modulus, (psi)   Not available

Hardness, (Knoop)   Not available

THERMAL PROPERTIES, (298°K)

Linear Expansion
Coeff., (°K)$^{-1}$ 16.75 x 10^{-6} (313°K)

Thermal Conductivity
(10^{-4} cal/(cm sec °K)   150

Specific Heat,
(cal/g)/°K   0.0479 (573°K)

OPTICAL PROPERTIES (298°K)

Dispersion Equation   Not available

Transmission Region, (External
Transmittance ≥10% with   2.0   mm.
thickness)   3.5 - 8.0μ

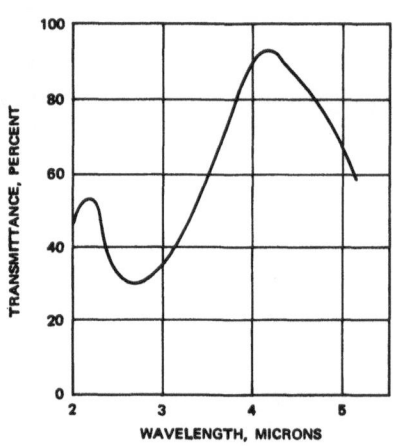

(REF. 1)

MATERIAL: <u>TELLURIUM</u>
(Polycrystalline
Film)

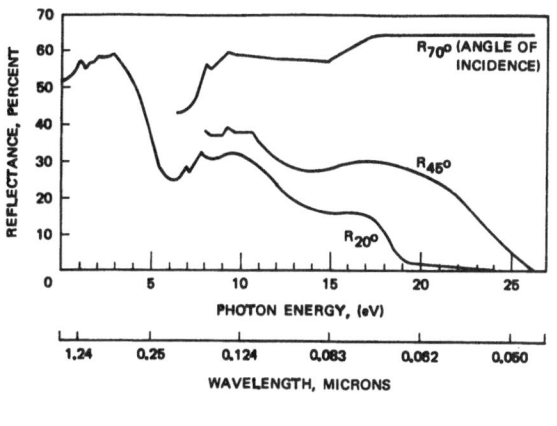

(REF. 2)

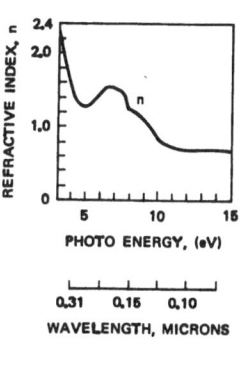

(REF. 2)

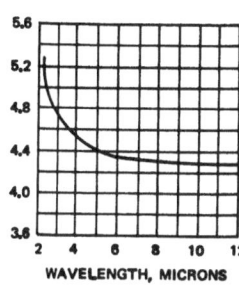

(REF. 1)

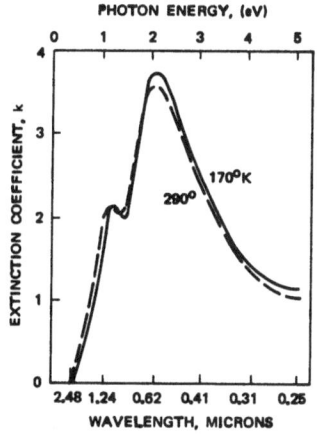

(REF. 3)

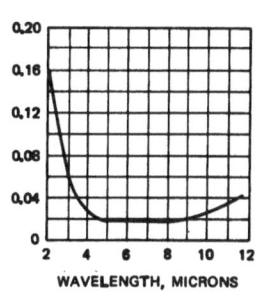

(REF. 1)

REFERENCES:

1. A.S. Valeer and M.A. Gisin, Optics and Spec., 19, 62-65, (1965).

2. J.D. Hayes, et al., J. Appl. Phys, 39, 5527-5532, (1968).

3. H. Keller and J. Stuke, Phys. Stat. Solidi, 8, 831-840, (1965).

TELLURIUM
(Single Crystal)

OPTICAL MATERIALS PROPERTIES
DATA SHEET

MATERIAL: TELLURIUM
(Single Crystal)

INTRODUCTION: This data sheet presents information information for single crystal tellurium.

PHYSICAL PROPERTIES, (298°K)

Density, (g/cm³)   6.25

Melting/Softening
Temp. (°K)   725

Solubility in Water,
(g./100 g. H_2O)   <0.005

MECHANICAL PROPERTIES, (298°K)

Young's Modulus, (psi)   Not available

Hardness, (Knoop)   18.4

THERMAL PROPERTIES*, (298°K)

Linear Expansion
Coeff., (°K)⁻¹   (-1.6, +27) x 10⁻⁶

Thermal Conductivity
(10⁻⁴ cal/(cm sec °K)   150 (320°K)

Specific Heat,
(cal/g)/°K   0.048

*E ∥ C and E ⊥ C axis respectively

OPTICAL PROPERTIES, (298°K)

Dispersion Equation   Not available

Transmission Region, (External
Transmittance ≥10% with 2.0 mm.
thickness)   3.5 - 8.0μ

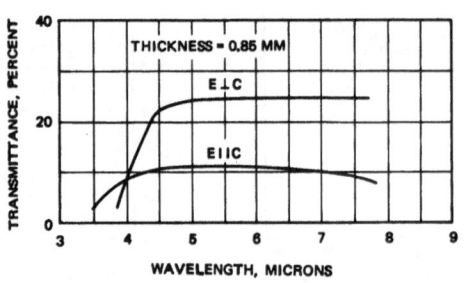

(REF. 1)

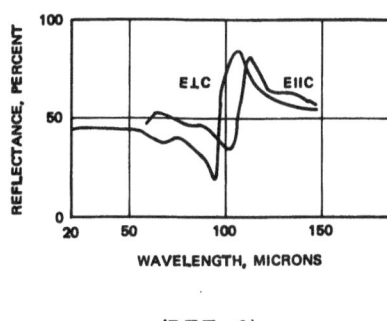

(REF. 2)

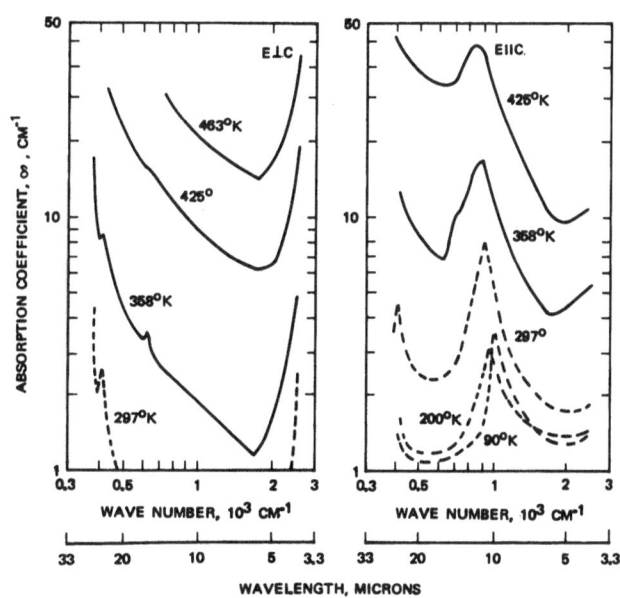

(REF. 3)

(REF. 4)

REFERENCES:

1. P. Billard, Acta Electronica, <u>6</u>, 75-169, (1962).

2. P. Grosse, et al, Solid State Comm., <u>5</u>, 99-100, (1967).

3. R.S. Caldwell and H.Y. Fan, Phys. Rev., <u>114</u>, 664-675, (1959).

4. H. Y. Fan, Reports on Progress in Physics, <u>19</u>, 107-155, (1956).

THALLIUM BROMOIODIDE

OPTICAL MATERIALS PROPERTIES
DATA SHEET

MATERIAL: THALLIUM
BROMOIODIDE
(KRS-5)

INTRODUCTION: This data sheet describes the properties of mixed crystal thallium bromoiodide (KRS-5).

PHYSICAL PROPERTIES, (298°K)

Density, (g/cm^3)   7.37

Melting/Softening
Temp. (°K)   688

Solubility in Water,
(g./100 g. H$_2$O)   0.05

THERMAL PROPERTIES, (298°K)

Linear Expansion
Coeff., (°K)$^{-1}$ 58 x 10^{-6}

Thermal Conductivity
(10^{-4} cal/(cm sec °K)   13

Specific Heat,
(cal/g)/°K   Not available

MECHANICAL PROPERTIES, (298°K)

Young's Modulus, (psi)   2.3 x 10^6

Hardness, (Knoop)   40.2

OPTICAL PROPERTIES, (298°K)

Dispersion Equation

$$n^2 - 1 = \sum_i \frac{K_j \lambda^2}{\lambda^2 - \lambda_j^2}$$

λ_j^2			K_j
λ_1^2	0.0225	K_1	1.8293958
λ_2^2	0.0625	K_2	1.6675593
λ_3^2	0.1225	K_3	1.1210424
λ_4^2	0.2025	K_4	0.04513366
λ_5^2	27089.737	K_5	12.380234

Transmission Region, (External
Transmittance ≥10% with   2.0   mm.
thickness)   0.6 - 40µ

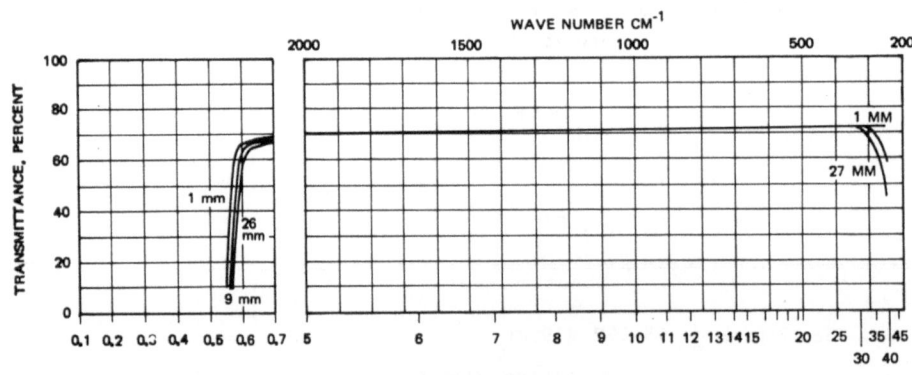

(REF. 1)

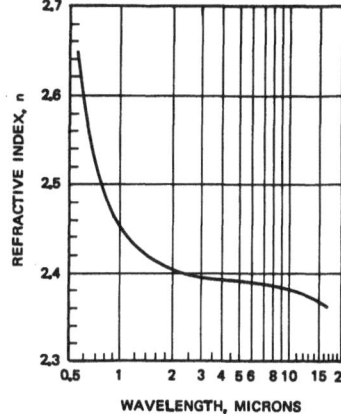

(REF. 3)

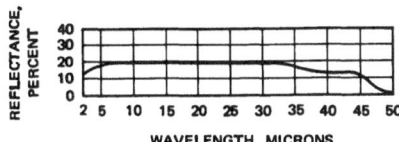

(REF. 2)

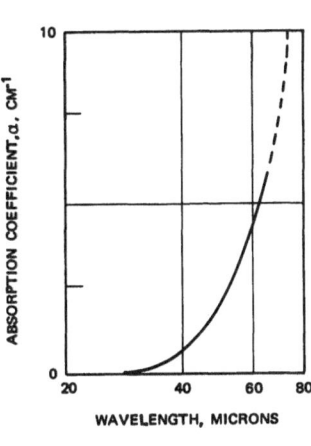

(REF. 4)

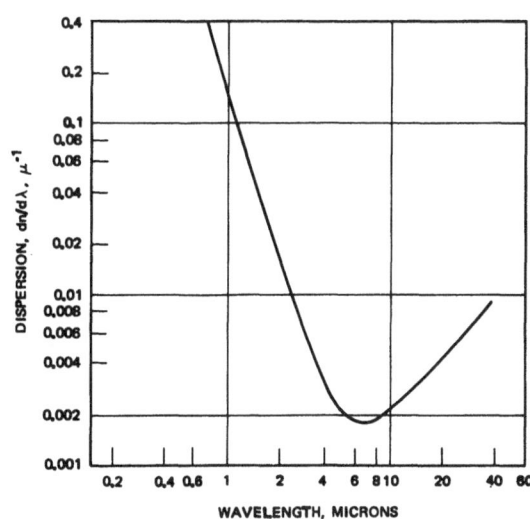

(REF. 5)

REFERENCES:

1. A. Smakula, et al, "Harshaw Optical Crystals," Harshaw Chemical Co., Cleveland, (1967).

2. D.E. McCarthy, Appl. Optics, 2, 591-595, (1963).

3. G. Joos, "F.I.A.T. Review of German Science, 1934-1946, The Physics of Solids," Part II, (1948).

4. A. Smakula, "Basic Properties of Optical Crystals with Special Reference to Infrared," Report No.PB111053, (1954).

5. A. Smakula, Opt. Acta., 9, 205-222, (1962).

THALLIUM CHLOROBROMIDE

<table>
<tr><td colspan="2" align="center">OPTICAL MATERIALS PROPERTIES
DATA SHEET</td><td>MATERIAL: THALLIUM
CHLORO-
BROMIDE
(KRS-6)</td></tr>
</table>

INTRODUCTION: This data sheet presents data for the mixed thallium chloride-thallium bromide crystal.

PHYSICAL PROPERTIES, (298°K)

Density, (g/cm³)　　7.192

Melting/Softening
Temp. (°K)　　697

Solubility in Water,
(g./100 g. H_2O)　　0.32 (293°K)

MECHANICAL PROPERTIES, (298°K)

Young's Modulus, (psi)　　3.0×10^6

Hardness, (Knoop)　　~30 (500 g)

THERMAL PROPERTIES, (298°K)

Linear Expansion
(Coeff., (°K)$^{-1}$　　50×10^{-6}

Thermal Conductivity
(10^{-4} cal/(cm sec °K)　　17.1 (329°K)

Specific Heat,
(cal/g)/°K　　0.0482

OPTICAL PROPERTIES, (298°K)

Dispersion Equation　　Not available

Transmission Region, (External
Transmittance ≥10% with　2.0　mm.
thickness)　　0.21 - 35µ

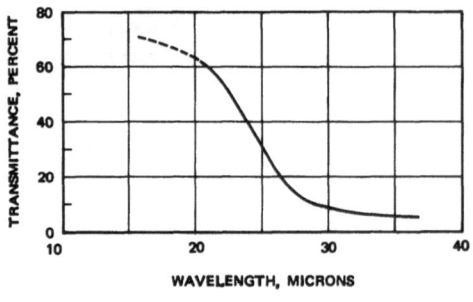

(REF. 1)

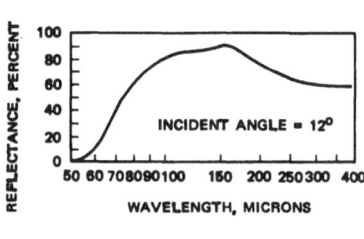

(REF. 2)

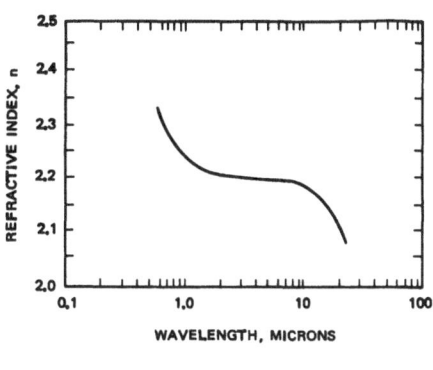

(REF. 3)

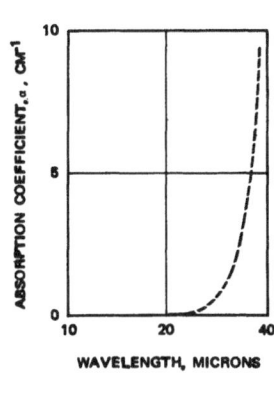

(REF. 4)

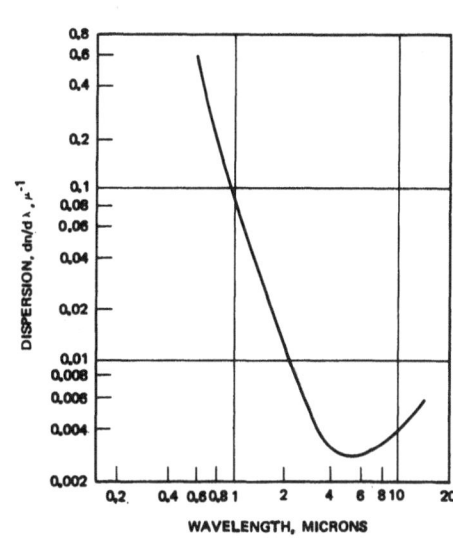

(REF. 4)

REFERENCES:

1. "American Institute of Physics Handbook," McGraw-Hill Book Co., New York, Second Ed., (1963).

2. A. Hadni, "Essentials of Modern Physics Applied to the Study of the Infrared," Pergamon Press, Oxford, (1967).

3. G. Hettner and G. Leisegang, Optik, <u>3</u>, 305-314, (1948).

4. A. Smakula, Opt. Acta, <u>9</u>, 205-222, (1962).

TITANIUM

OPTICAL MATERIALS PROPERTIES DATA SHEET

MATERIAL: TITANIUM

INTRODUCTION: This data sheet contains information on bulk titanium, except for film transmittance and reflectance data.

PHYSICAL PROPERTIES, (298°K)

Density, (g/cm^3)   4.50

Melting/Softening Temp. (°K)   2085

Solubility in Water, (g./100 g. H_2O)   <0.01

MECHANICAL PROPERTIES, (298°K)

Young's Modulus, (psi) 15×10^6

Hardness, (Brinell)   55 - 95

THERMAL PROPERTIES, (298°K)

Linear Expansion Coeff., (°K)$^{-1}$ 8.4×10^{-6}

Thermal Conductivity (10^{-4} cal/(cm sec °K)   490

Specific Heat, (cal/g)/°K   0.126

OPTICAL PROPERTIES, (298°K)

Dispersion Equation   Not available

Transmission Region, (External Transmittance ≥10% with_____mm. thickness)   Not available

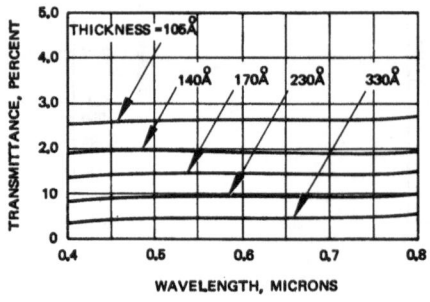

(REF. 1)

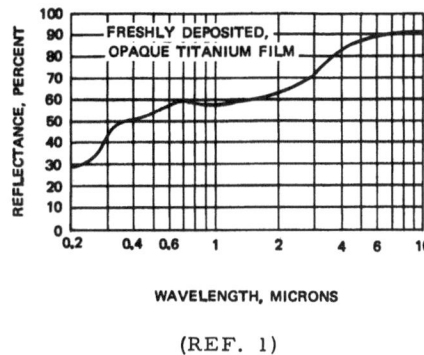

(REF. 1)

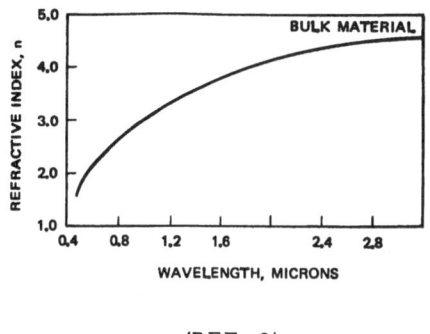

(REF. 2)

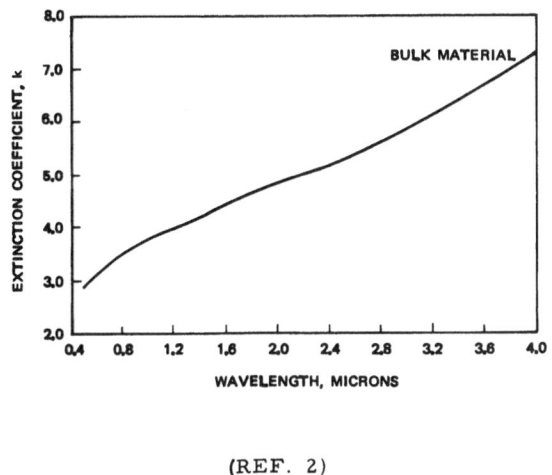

(REF. 2)

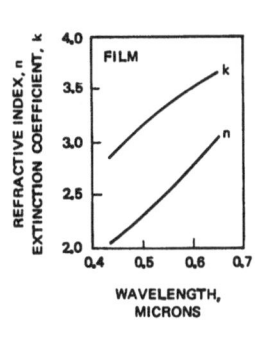

(REF. 1)

REFERENCES:

1. G. Hass and A.P. Bradford, J. Opt. Soc. Am., 47, 125-129, (1957).

2. M.M. Kirillova and B.A. Charikov, Phys. of Metals and Metall., 15, 138-139, (1963).

TITANIUM DIOXIDE

OPTICAL MATERIALS PROPERTIES
DATA SHEET

MATERIAL: <u>TITANIUM</u>
<u>DIOXIDE</u>
<u>(Rutile)</u>

INTRODUCTION: <u>This data sheet contains information for single crystal stoichiometric titanium</u>
<u>dioxide.</u>

PHYSICAL PROPERTIES, (298°K)

Density, (g/cm^3) ___4.26___

Melting/Softening
Temp. (°K) ___2093___

Solubility in Water,
(g./100 g. H$_2$O ___<0.001___

MECHANICAL PROPERTIES, (298°K)

Young's Modulus, (psi) ___Not available___

Hardness, (Knoop) ___879 (500 g)___

THERMAL PROPERTIES*, (298°K)

Linear Expansion
Coeff., (°K) ___(9.19 - 7.14) x 10^{-6}___

Thermal Conductivity ___300 ∥ C at 309°K,___
(10^{-4} cal/(cm sec °K) ___210 ⊥ C at 317°K___

Specific Heat,
(cal/g)/°K ___0.17___

*Double values for E ∥ C and E ⊥ C respectively.

OPTICAL PROPERTIES, (298°K)

Dispersion Equation

Ordinary ray:

$$n^2 = 5.913 + 2.441 \times 10^7/(\lambda^2 - 0.803 \times 10^7)$$

Extraordinary ray:

$$n^2 = 7.197 + 3.322 \times 10^7/(\lambda^2 - 0.843 \times 10^7)$$

where $\lambda = \overset{\circ}{A}$

Transmission Region, (External
Transmittance ≥10% with ___2.0___ mm.
thickness) ___0.43 - 6.2µ___

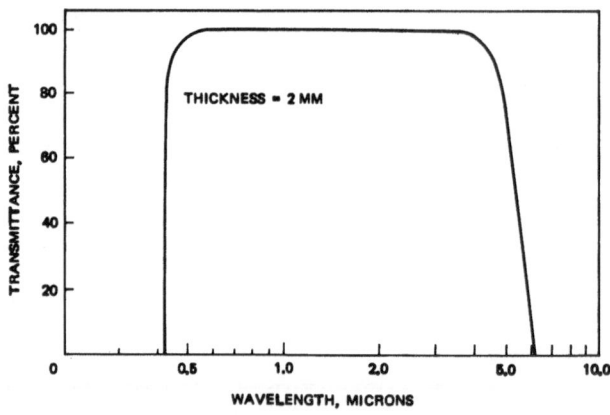

(REF. 1)

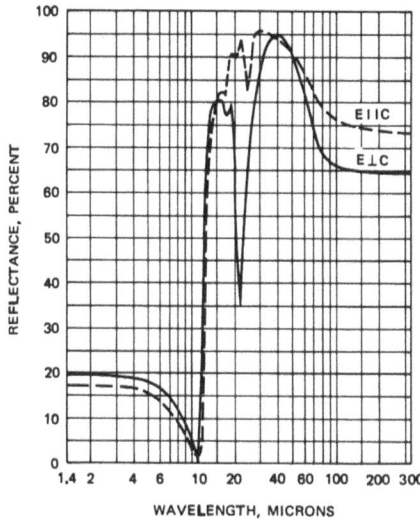

(REF. 2)

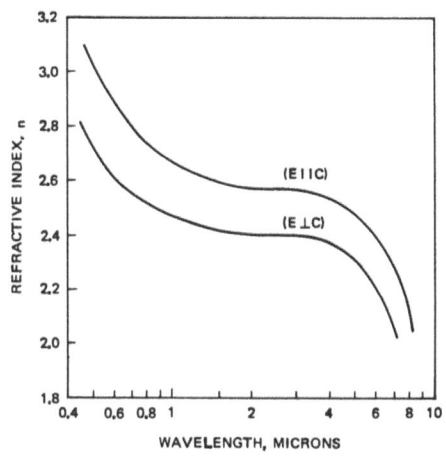

(REF. 3)

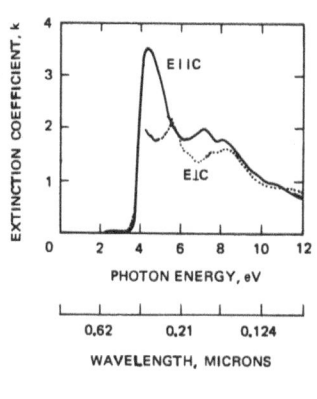

(REF. 4)

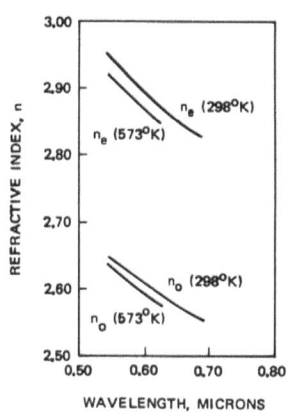

(REF. 5)

REFERENCES:

1. M. D. Beals and L. Merker, Materials in Design Eng., <u>51</u>, No. 1, 12-13, (1960).

2. D. C. Cronemeyer, Phys. Rev., <u>87</u>, 876-886, (1952).

3. D. C. Cronemeyer and M. A. Gilleo, Phys. Rev., <u>82</u>, 975-976, (1951).

4. M. Cardona and G. Harbeke, Phys. Rev. <u>137</u>, A1467-A1476, (1965).

5. A. Schroeder, Z. Kristallog, <u>67</u>, 485-542, (1928).

ZINC SELENIDE

OPTICAL MATERIALS PROPERTIES
DATA SHEET

MATERIAL: ZINC SELENIDE
(Cubic)

INTRODUCTION: This data sheet contains data for cubic zinc selenide.

PHYSICAL PROPERTIES, (298°K)

Density, (g/cm^3)   5.651

Melting/Softening
Temp. (°K)   1373

Solubility in Water,
(g./100 g. H_2O)   <0.001

MECHANICAL PROPERTIES, (298°K)

Young's Modulus, (psi)   Not available

Hardness, (Mohs)   3 - 4

THERMAL PROPERTIES, (298°K)

Linear Expansion
Coeff., (°K)$^{-1}$ 7×10^{-6}

Thermal Conductivity
(10^{-4} cal/(cm sec °K)   290

Specific Heat,
(cal/g)/°K   0.016

OPTICAL PROPERTIES, (298°K)

Dispersion Equation   Not available

Transmission Region, (External
Transmittance ≥10% with 2.0 mm.
thickness)   ~0.5 - $\overline{22}$μ

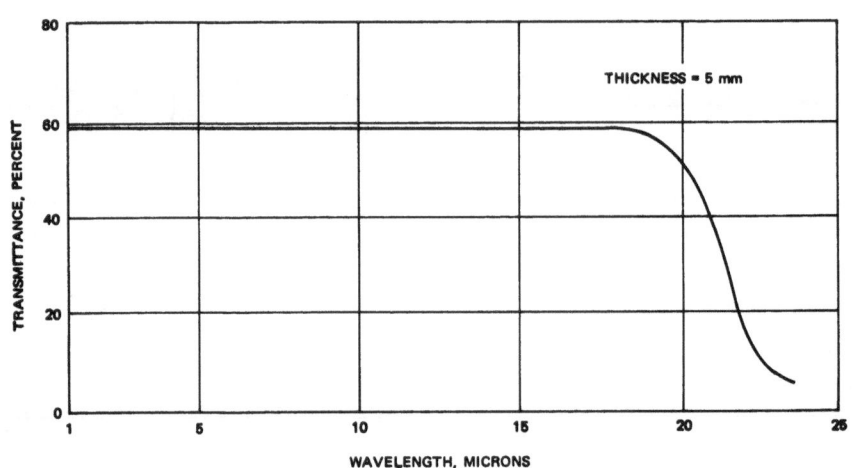

(REF. 1)

98

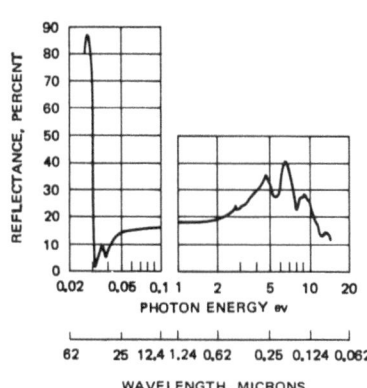

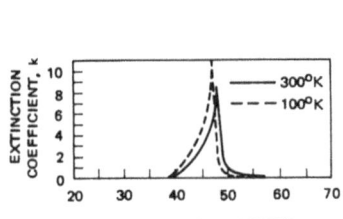

(REF. 2)

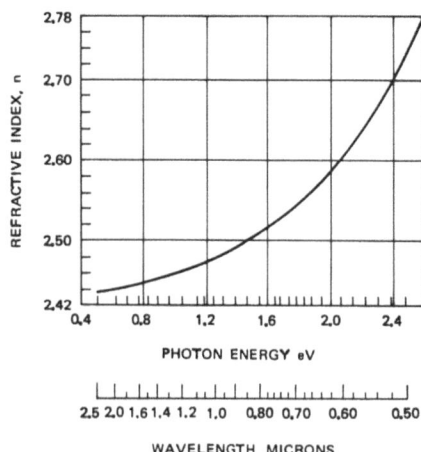

(REF. 3)

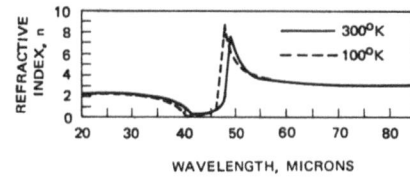

(REF. 4)

(REF. 4)

REFERENCES:

1. K.K. Dubenskiy, Soviet J. of Opt. Tech., 36, 118-121, (1969).

2. M. Aven, et al, J. Appl. Phys. Suppl., 32, 2261-2265, (1961).

3. D.T.F. Marple, J. Appl. Phys., 35, 539-542, (1964).

4. A. Manabe, et al, Japan, J. Appl. Phys., 6, 593-600, (1967).

ZINC SULFIDE

OPTICAL MATERIALS PROPERTIES
DATA SHEET

MATERIAL: ZINC SULFIDE
(Cubic)

INTRODUCTION: This data sheet presents information for single crystal cubic zinc sulfide.

PHYSICAL PROPERTIES, (298°K)

Density, (g/cm^3)   4.09

Melting/Softening
Temp. (°K)   1293

Solubility in Water,
(g./100 g. H_2O 6.5 x 10^{-5} (299°K)

MECHANICAL PROPERTIES, (298°K)

Young's Modulus, (psi)   1.65 x 10^6

Hardness, (Mohs)   3.5 - 4

THERMAL PROPERTIES, (298°K)

Linear Expansion
Coeff., (°K)-1   6.14 x 10^{-6} (273°K)

Thermal Conductivity
(10^{-4} cal/(cm sec °K)   635

Specific Heat,
(cal/g)/°K   0.116

OPTICAL PROPERTIES, (298°K)

Dispersion Equation

$$n^2 = 5.164 + 1.208 \times 10^7/(\lambda^2 - 0.732 \times 10^7)$$

where $\lambda = \overset{\circ}{A}$ '

Transmission Region, (External
Transmittance ≥10% with   0.62   mm.
thickness)   ~0.6 - 15.6μ

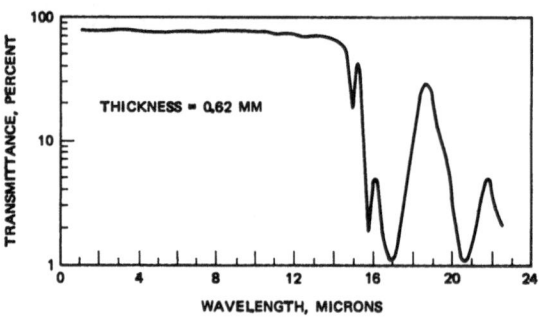

THICKNESS = 0.62 MM

(REF. 1)

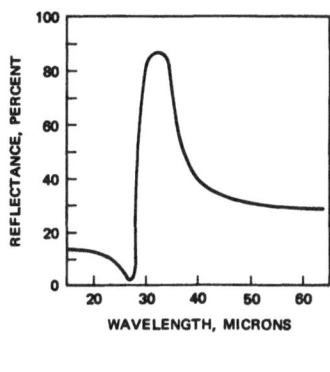

(REF. 2)

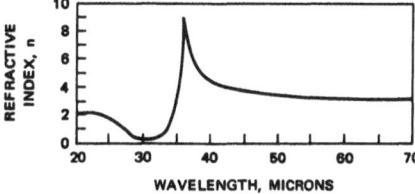

(REF. 3)

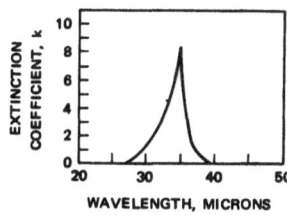

(REF. 2)

(REF. 2)

REFERENCES:

1. T. Deutsch, "Int. Conf. on the Phys. of Semiconductors, Proc. Exeter, 1962," A.C. Stickland, Ed., Inst. of Phys. & Phys. Soc., London, 505-512, (1962).

2. A. Manabe, et al, Japan. J. Appl. Phys., 6, 593-600, (1967).

3. S.J. Czyzak, et al, U.S. Government Report No. AD-143919, (1957).

APPENDIX A
WAVELENGTH CONVERSION FACTORS

(1) CONVERSION FACTORS FROM
ELECTRONS VOLTS, (eV) to MICRONS (µ)

eV	µ	eV	µ	eV	µ
0.10	12.39	1.17	1.06	4.4	0.28
0.12	10.33	1.20	1.03	4.5	0.28
0.14	8.86	1.25	0.99	4.6	0.27
0.16	7.75	1.35	0.92	4.7	0.26
0.18	6.89	1.40	0.89	4.8	0.26
0.20	6.20	1.45	0.86	4.9	0.25
0.22	5.64	1.50	0.83	5.0	0.25
0.24	5.17	1.55	0.80	5.1	0.24
0.26	4.77	1.60	0.78	5.2	0.24
0.28	4.43	1.65	0.75	5.3	0.23
0.30	4.13	1.70	0.73	5.4	0.23
0.32	3.87	1.75	0.71	5.5	0.23
0.34	3.65	1.80	0.69	5.6	0.22
0.36	3.44	1.85	0.67	5.7	0.22
0.38	3.26	1.90	0.65	5.8	0.21
0.40	3.10	1.95	0.64	5.9	0.21
0.42	2.95	2.0	0.62	6.0	0.21
0.44	2.82	2.1	0.59	7.0	0.18
0.46	2.70	2.2	0.56	8.0	0.16
0.48	2.59	2.3	0.54	9.0	0.14
0.50	2.48	2.4	0.52	10.0	0.12
0.53	2.34	2.5	0.50		
0.56	2.21	2.6	0.48		
0.59	2.10	2.7	0.46		
0.62	2.00	2.8	0.44		
0.65	1.91	2.9	0.43		
0.68	1.82	3.0	0.41		
0.71	1.75	3.1	0.40		
0.74	1.68	3.2	0.39		
0.77	1.61	3.3	0.38		
0.80	1.55	3.4	0.36		
0.83	1.49	3.5	0.35		
0.86	1.44	3.6	0.34		
0.89	1.39	3.7	0.34		
0.92	1.35	3.8	0.33		
0.95	1.30	3.9	0.32		
0.98	1.26	4.0	0.31		
1.11	1.12	4.1	0.30		
1.14	1.09	4.2	0.30		
		4.3	0.29		

(2) CONVERSION FACTORS FROM WAVENUMBER, (cm^{-1}) to MICRONS, (μ)

Wavenumber cm^{-1}	μ	Wavenumber cm^{-1}	μ	Wavenumber cm^{-1}	μ
100	100	800	12.5	6500	1.54
110	90.9	825	12.12	7000	1.43
120	83.3	850	11.76	7500	1.33
130	76.9	900	11.11	8000	1.25
140	71.4	925	10.81	8500	1.18
150	66.7	950	10.53	9000	1.11
160	62.5	975	10.26	9500	1.05
170	58.8	1000	10.00	1×10^{4}	1.00
180	55.6	1050	9.95	1.1	0.91
190	52.6	1100	9.09	1.2	0.83
200	50.0	1150	8.70	1.3	0.77
220	45.5	1200	8.33	1.4	0.72
240	41.7	1250	8.00	1.5	0.67
260	38.5	1300	7.69	1.6	0.62
280	35.7	1350	7.45	1.7	0.59
300	33.3	1400	7.14	1.8	0.56
320	31.2	1450	6.90	1.9	0.53
340	29.4	1500	6.67	2.00	0.50
360	27.8	1550	6.45	2.1	0.48
380	26.3	1600	6.25	2.2	0.46
400	25.0	1650	6.06	2.3	0.42
420	23.8	1700	5.88	2.4	0.38
440	22.7	1750	5.71	2.5	0.37
460	21.7	1800	5.56	3.0	0.33
480	20.8	1850	5.41	3.5	0.28
500	20.0	1900	5.26	4.0	0.25
520	19.2	1950	5.13	4.5	0.22
540	18.5	2000	5.00	5.0	0.20
560	17.9	2200	4.55	6.0	0.17
580	17.2	2400	4.17	7.0	0.14
600	16.7	2600	3.85	8.0	0.12
620	16.1	2800	3.57	9.0	0.11
640	15.6	3000	3.33	1×10^{5}	0.10
660	15.2	3500	2.85		
680	14.7	4000	2.50		
700	14.3	4500	2.22		
725	13.8	5000	2.00		
750	13.3	5500	1.82		
775	12.9	6000	1.67		

APPENDIX B

GLOSSARY OF OPTICAL TERMS

ABSORPTION COEFFICIENT, (α) - defined by the equation:

$$I_x = I_o \, e^{-\alpha x}$$

where I_o = incident radiation intensity

                  I_x = transmitted radiation intensity at distance of x cm into the material.

                  α = absorption coefficient, cm^{-1}

BIREFRINGENCE - division of incident light by optically anisotropic materials into
   (Double   two component which are diffracted in different directions.
Refraction)

DISPERSION, ($dn/d\lambda$) - the derivative of the refractive index with respect to wavelength,

     where: λ = wavelength, microns.

EXTINCTION COEFFICIENT, (k) - defined by the equation:

$$k = \frac{\alpha\lambda}{4\pi}$$

      where λ = wavelength in cm
                 α = absorption coefficient
                 k = extinction coefficient, dimensionless

EXTRAORDINARY RAY - wave traveling with a velocity which depends on the relation between its direction and the optic axis, being polarized with its electric vector parallel ($\parallel$) to the plane containing the optic axis and the direction of propagation. (See also birefringence)

ORDINARY RAY - wave traveling with a velocity independent of the direction of propagation, being polarized with its electric vector perpendicular ($\perp$) to the direction of the optic axis and the direction of propagation. (See also birefringence)

REFLECTANCE, (R)- The percentage of the incident light reflected by a surface.

REFRACTIVE INDEX, (n) - the ratio of the phase velocity of light in vacuum to the phase velocity in the medium.

TRANSMITTANCE, (T) - the percentage of incident light transmitted through the material and observed. (also called external transmittance)